GODLINESS:
Building Your Own Cosmology

by

Charles W Heath, Master of the Cosmos (ret)

ISBN: 1-4033-7479-1 (e-book)
ISBN: 1-4033-7480-5 (Paperback)

Library of Congress Control Number: 2002094467

This book is printed on acid free paper.

Printed in the United States of America
Bloomington, IN

1stBooks – rev. 03/19/03

<u>Preface</u>

Books combining gospel theology and junk science have become more and more common. It seemed a nice idea to reverse the trend. This little tome blends junk theology with real science. At least it is real science as far as it goes. The mix of deities and donuts sets out a pattern for a universe that, while huge, is not infinitely large in the amount of matter and energy it contains or the space it occupies. That should make an easier pill to swallow than the *dictum* that our universe is infinitely massive.

Do attempt to pick up on the nonsense. Some has been scattered around on purpose. Try to spot the baloney included by mistake. A lot of work I have not done is mentioned in Chapter 7, "Loose ends." Give it a go.

Engineers, physicists, astronomers and similar oddballs may enjoy this book. I expect the many numbers in Chapter 5 plus – gasp! – algebra and calculus in the Appendix could put off a few of our general readers.

Charles Heath, Houston, August 2002

Table of Contents

To be a god

Mel Brooks has said, more than once, "It's good to be a king." I have never been a king, but I certainly agree in principle. Speaking for myself, I paraphrase the noble Mel and say, "It's good to be a god."

Let's get something straight from the beginning: you must never say nor imply, "I am the God." The God has too much responsibility for one thing. Also if you go about claiming to be the God, you are sure to meet a lot of rude policemen and psychiatric nurses – life is complicated enough without that. Of course, some people have made a good living out of being a spokesperson for the God, but that is very different from what I have in mind. Self-satisfaction is what makes it good to be a god. You may impress a few others with your godlike demeanor, but not your spouse or nineteen-year-old child. However, you can rely on your own opinion – your own good opinion – almost always and always increasingly.

GODLINESS: _Chapter One_

I have thought of myself as a god for a great many years, but was judicious enough not to say it aloud. I was a covert god. Then, in the early nineties, I decided to become a cruising live-aboard and sail around part of the world. To that end, I bought an auxiliary sailing yacht and named it _Cosmos._ As Captain and Owner I was Master of that vessel. I began to sign my name:

"Charles... Master of the Cosmos"
Eureka! I became an overt god just like that. Out of the closet at last. I have since sold the boat and added "(ret)" to my title.

Let me interject here that the proper honorific of address for such a god is "Magnifico." It is proper because I say so. I like it because it has nuances of sham, flimflam and obsequiousness. It derives from Latin and works well in English and the romance languages. It sort of means "big-shot", or that other phrase you get by transposing the vowels. Since those great many years ago, whenever people ask me what I prefer to be called, I answer, "Magnifico." They may use the form at times, yet only one person – a woman, not my wife – ever so addressed me consistently. Maybe they merely wonder if I want to be called Charlie or Chuck, but I don't much care for either nickname.

You may have noticed that I use the first person singular when referring to myself. Some gods use the plural form, "we", instead. That is entirely acceptable. After all, individual kings and newspaper editors have called themselves "we" for centuries. It's a matter of personal taste.

A god, according to my definition, is any being who can do impossible things while fully understanding that what is being done is impossible and not of this real world. To do magic without understanding its unreality is not godlike. A common woods nymph might turn cucumbers into sunlight by making time run backwards. But if she doesn't realize that's not possible, she's not a god. Being a god is often confined to working with one's mind.

I became omniscient when I turned into a god. It's too bad that, like all great gifts,

GODLINESS: Chapter One

omniscience came with a catch. I now have a rotten memory. So although I know everything, I can remember only the tiniest fraction of it all – a fraction that grows tinier every day.

My dictionary suggests that a god is believed, among other things, to require human worship. What nonsense! I would be annoyed and embarrassed if anyone tried to worship me. Maybe if I were Greek and younger, it would be nice to be worshiped by some of the more nubile females. I don't know all the gods. Those I do know also dislike being worshiped except, perhaps, by a dog.

In this little tract, I am going to set forth a few of my own biases, a general precept and a complex cosmology to illustrate that general precept. If you are able to stick with it and grasp the cosmology, you will become a god. You don't have to buy a yacht. My current title, Master of the Cosmos (ret), is not the real reason I am a god. I became a god the same way you or anyone else can: by doing at least one piece of magic while fully aware that it was not possible. The magic need not involve crashing mountains together, bending spoons or anything else physical, which must come as a relief. No, it can be done entirely within your own head as an imaginative visualization.

There are other paths toward godhood besides cosmology. Two examples: 1) Beethoven composed his Ninth Symphony, a secular work. Its magic reconfirmed what was already known. 2) The Monty Python group made a film called *The Meaning of Life* that may have exalted some of them. It's hard to tell which.

Cosmology was once filed under religion. Nowadays it's an offshoot of physics, a branch of natural science. That's a place where I can find peace and comfort, and hear some sweet music.

Perhaps the nicest thing about being a god is to gain that smug sense of superiority over the non-gods one meets. Superiority promotes a calm, benign attitude of *noblesse oblige.* Although there are a few jealous gods, the benignity tends to overshadow

GODLINESS: Chapter One

petty quarrels between them. It would be immodest to claim that this little essay will introduce a new golden age of courtesy to our society. Nevertheless, I think it will.

Along time ago I went to school and became a Chemical Engineer. I got a solid grounding in math, thermodynamics, physical chemistry and other sexy subjects pertinent to physics. Later on, I taught myself Relativity Theory from a collection of Albert Einstein's writings called *Relativity: The Special and General Theory.* Uncle Bertie became my personal guru. I felt I knew everything.

Clearly I either did not know everything or had forgotten most of it, but I felt that way anyhow. My grasp of spatial relations was good and I sensed that I knew where everything was. I don't mean everything in my house; I mean everything in the universe. I still feel that way except briefly after deep research into the effects of Gin or Vodka. That's when I barely know where anything is.

Whenever a new fact or idea or allegation turns up, it either does or does not fit into and reinforce my world-view. If it seems not to fit in, I ponder at length. Then it either gets chucked out or tabled for a future decision or I alter my world-view. I am not ashamed to make a change; I just let on that I always felt that way. My world-view is what I call "commonsense."

* * *

A scientist named George Gamow wrote a book in the early fifties called _The Creation of the Universe._ It was meant for laymen and described the Big Bang theory in a good deal of detail. Much has been written about Big Bang theory, but it was Gamow's book that I happened to be reading in December of 1971 and January of 1972.

Gamow postulated that the universe contained an infinite amount of "matter or energy or both" – that phrase can be implicit in either word alone. He said that the matter/energy of the universe was extremely dense a few billions of years ago relative to its present density. That density in some way corresponded to what might be called "critical mass" and all the matter/energy exploded simultaneously in a Big Bang. That explosion was equivalent to an "open" expansion that is still underway and will continue forever. Atomic species as we know them now were largely formed during the first few hours. The concept of time previous to the Big Bang is meaningless.

From Gamow's book, it seems that time started from zero at the instant of the Big Bang, and hence time is not infinite into the past. Yet time is infinite into the future, however meaningless that future may become as the universe unwinds into a dead cinder approaching absolute zero temperature and density. The amount of matter/energy in the universe is infinite, and the size of the universe is infinite.

Gamow describes clearly how an infinite amount of matter/energy does not "fill" an infinite amount of space; there's always room for more, even an infinite amount more. He also shows how space can be boundless while finite; much the same way as the perimeter of a wheel or the surface of a sphere describes unbounded but finite distance (one-dimensional space) or area (two-dimensional space).

GODLINESS: Chapter Two

Now, I like parts of this story very much, but other parts put a wire edge on my teeth. The good stuff includes the Big Bang itself, the cooking of atomic species, the ongoing expansion and, of course, space without a boundary wall anywhere. The troublesome bits are applying the concept of infinity to the amount of matter/energy in the universe and denying the same concept to time in the past. These two thoughts run counter to my notions of commonsense. Also, the broad meaning of "simultaneous" was shot down by Special Relativity.

I might add that it seems too, too dismal for a universe to exist forever in a run-down comatose state. It would be far cheerier if our universe, like Tigger, kept bouncing. I am told that Uncle Bertie once preferred the idea of a cyclical universe. He didn't push it, maybe because he had better fish to fry.

Estimates of the age and average density of the universe are just that – estimates. Increasing those estimates by not so very much could cause the universe to "close." That is, a small increase could allow gravity to overcome the expansion, leading to contraction and a Big Crunch that might turn into another Big Bang. For the universe to be poised so giddily between "open" and "closed" makes me nervous.

Infinity is a concept in math. As a number, it is larger than any number that can be imagined even if that number is multiplied by itself, raised to any power and made factorial. It can be handled in math by a set of simple rules. Addition to or subtraction from infinity is meaningless; the result is infinity. The same is true of the other arithmetic processes. If infinity is used as a devisor, the result is an approach to zero as in "infinitesimally small." That latter phrase was involved in the development of calculus. Infinity is a concept often used as a limiter, for example, "A spaceship traveling near the speed of light cannot accelerate past light speed even if it could expend an infinite amount of energy in the attempt."

Nothing that I know of the material world is infinite. Infinitely dense, infinitely hot, infinitely hard, infinitely fast are properties that do not seem to exist. Infinity is a concept. It can be applied with ease to other concepts like time and space.

Size appears to be defined by vectored distances between objects that are, or may be, contained within a space. When Gamow said the size of the universe is infinite, I believe he merely meant that it had to be infinite to contain the infinite amount of matter/energy he had postulated. The possibility of infinite space of itself does not annoy me.

Somebody once said, "The purpose of time is to keep everything from happening at once." Time often appears to me as a process of counting repetitions – day, night, day etc – and linking all other events to various stages of the count. Obvious reality is the present effect of past cause. Equally certain is that present cause results in future effect. My basic sense of the zero point in time is that zero is right now. From right now, I can count forward into the future or backward into the past. This progress through time calls for an infinite future and an infinite past; any boundary would be most arbitrary. Relativity Theory treats time as a fourth dimension, mathematically the same as the other three familiar dimensions. That's OK by me, if needed; but it's often easier to visualize our usual world in a three-dimensional volume, with time as a distinct kind of extra dimension.

Perhaps when Gamow said that time previous to the Big Bang was meaningless he merely meant that no artifact or data could survive the intense heat, pressure and turmoil of that moment. Fair enough, but ideas can pierce such armor. I like the idea that a Big Crunch immediately preceded the Big Bang.

So I've got a prejudice, a bias, an attitude. I don't want to accept a cosmology that calls for a universe of infinite size, containing an infinite amount of matter/energy and with a limited past existence. Neither do I like a cosmology that has its universe dribbling away forever with an endless whimper.

* * *

<u>GODLINESS</u>: *Chapter Two*

Around about here, an original insight suddenly sprang forth! – original to me, at any rate. I put it that *"A cogent theory can be found which supports any set of non-contradictory postulates."* This sounded like a potentially useful bit of wisdom. Alas, I did not have the ghost of an idea how to prove its truth, nor do I today.

If I could not prove, at least I could demonstrate by building a cosmology that met my personal biases while not conflicting with any of Gamow's evidence. This I did by working up a "partial cosmology of the Monodim world" as an analog model of our own universe. It came out pretty well and suggested by analogy that our own universe exists in five spatial dimensions.

Now any reader still awake is bound to ask why am I resurrecting this thirty-year old dodo? There are several reasons. Among them: I have expanded my "partial cosmology..." to a "general cosmology..." and dropped some thoughts that didn't stand up too straight. I claimed I would show you how to become a god. I have a hidden purpose. And I want to toss my *"A cogent theory..."* into the ring for you to kick around.

That *"cogent theory"* seems to have a lot to do with semantics. It brings to mind some old sayings like: "lies, damned lies and statistics," "you can fool all the people some of the time" and "Figures don't lie, but liars can figure." Above all, it makes me wonder about the difference between actual reality and perceived reality. I'm sure there is a disparity, but don't tell me the difference is infinite.

There was a time when most people thought the earth was flat. Travelers feared they might fall off if they got too close to the edge. That difference between reality and our perceived version is less than it was ten thousand years ago. It is less than it was a thousand years ago, less than it was a century ago. We could not detect any shrinkage in an infinite difference. Knowledge could have no effect on such a gulf.

So we keep trying and we keep gaining ground. We don't know when or where it will end, but the journey is good fun.

* * *

More and more I hear the statement, "Anything is possible." This is said with a smirk that suggests the speaker is imparting great wisdom. I can think up things that are not possible and so can you. I watch reruns of *Star Trek: The Next Generation* and enjoy them. It is well-made and entertaining fiction, loaded with impossibilities. For a vessel or its systems of communication to exceed light-speed is surely impossible under Relativity Theory. Relativity may some day be extended in its turn just as it has extended Newton's Laws, but there is no sign of that happening as yet. The tractor beam, the transporter and the replicator are described so vaguely that their impossibility is hard to prove. Lots of things are merely absurd. A crewman does original research to save everyone's bacon at the last minute in many episodes. Out of six to nine hundred on board, less than a dozen adults and one teenager run it all. The captain spends much of his time on the bridge telling the helmsperson which way to steer and how fast to go. It's good-natured and I like it partly because of the silly stuff.

Of course, people often say something is impossible when they mean it's difficult, or they don't know how or don't want to do it. The word is misused. It does not follow that, "Anything is possible."

Impossibilities permit simple rules for complex results. They are as useful to science and technology as their inverse. To work out a material balance and an energy balance is one of the great tools of chemical engineering. Energy balance is basic to thermodynamics, which links heat energy and mechanical energy and bared the secrets of the steam engine. These balances embody the simple yet profound idea that neither matter nor energy can be created or destroyed. We depend on the impossibility of creating or destroying matter/energy. For a dynamic system in equilibrium, matter input = matter output, and energy input =

energy output. From Relativity we know that energy and matter are related by the famous formula, $E = mc^2$. In <u>almost all</u> engineering, we can ignore Relativity and just use the old concepts of separate matter and energy. In cosmology, we need Relativity and any other help we can get.

Tools

You need some background know-ledge to follow the next three chapters. That would be true even for reading *Pulp Fiction*. But physics is a realm that many people find mysterious and not much fun when compared to, say, rock music. The fun part is personal; I think physics is fun; you may not. Mystery is another story. Rock music is mysterious in that no one knows before the fact which piece or which band, out of the many that try, will shoot to the top. Physics is not mysterious at all, because it is sharply defined. You have to know the definitions and the jargon; they derive from what has been worked out earlier, and may be in the language of math.

It would be nice if you knew math through calculus, including analytic geometry, which is a fusion of algebra and Euclidean geometry. But a small remnant of algebra is all you really need. I will toss in a few, very few, formulas here and there and put some more into an appendix. I'm not going to teach any math.

GODLINESS: Chapter Three

It would be good if you knew Relativity Theory and, of course, Newton's Laws, which are a limited case within Relativity Theory – a case where the effects of high velocities and gravitational field strengths are negligible. There are plenty of self-teaching books on the Theory; Uncle Bertie's seems best to me. I'm going to make lots of statements from Relativity. If you don't know the Theory, you will have to take my assertions on faith. Is it unseemly for a god to accept things on faith? The answer would appear to be "no" although it makes me uncomfortable to say it. Perhaps a god is diminished but not disqualified thereby.

You must understand how basic navigation works on the surface of the Earth. I insist that you refer to a globe; an ordinary 12" globe is fine, but not included. If you don't have a globe or if yours is too old and beat up, you can buy one at Wal-Mart for about twelve bucks around the time the kids are going back to school. My last purchase was off-season and cost twenty dollars somewhere else. We deal with flat maps so much that our sense of geography tends to get distorted, and a globe points us back toward reality. A globe is a handy thing for lots of reasons aside from its use in cosmology. Mine also does duty as a high-frequency sound reflector in the audio section of my home theater.

I will require you to visualize a spherical surface in space, which has poles, an equator, parallels of latitude and meridians of longitude – all with the same layout and numbering system found on your globe. This will help me communicate with your imagination.

While sailing around the Caribbean, I kept a sharp eye on the water looking for lines of latitude and longitude during the daylight hours. I only spotted one in all that time. It was a dim yellowish group of frayed noodle-like things running north and south, a presumed meridian of longitude. I could not read anything on it and had to refer to my GPS receiver for its number. These lines have been so worn out over the eons since their creation that they have become imaginary for all practical purposes. If Columbus, Magellan, Nelson and their sub-captains could find their way around using imaginary lines, so can you. You won't even have to relate those lines to land and water, just to an imaginary universe.

When I find some cute pieces of clip art that seem to fit, I may stick them in here and there. But don't expect any fancy video to show you what you are supposed to visualize. A creative, dancing animation with music and a plummy voice-over narrative might seem very nice on the surface, but it would stifle your ability to envision on your own. We are so used to being spoon-fed with entertainment that the term "couch potato" has taken on a real meaning. It's a one-way street that numbs the mind. It kills imagination as it extends the waistline.

You may take these last remarks as sour grapes. I have neither the skill, will nor spirit to do a good animation myself, nor a desire to pay someone else for the job. Be content with a globe.

In this list of minimal tools needed, I have neglected the most important of all: that is, motivation. Your motives may be any or all of: 1) You want to become a god. 2) You want to learn how to build your own cosmology. 3) You want to find out about, argue with or modify my cosmology. 4) You just want to exercise your visualization skills.

That's it. Now dig in.

GODLINESS: Chapter Four

Let's set up an imaginary universe in one single spatial dimension. This is a mono-dimensional universe, having only distance for its space, but with time as well. It exists in a line, as a line segment of definite length. It's a big segment, some 66 billion light-years long! We do not want any clumsy boundaries in this universe, so we join the two ends of the line segment together to form a circle 21 billion light-years in diameter.

We do want this imaginary universe to resemble our own as much as possible; so we postulate that it contains matter/energy, with temperature, light, gravity and other qualities working much as they do in our world. It even has people, whom I call the Monodim, after <u>mono-dim</u>ensional space. This noun is both singular and plural – like sheep. The Monodim are intelligent beings, much as we are. They may not be as good-looking as Harry Hirsute or Pfanny Pfeiffer but, then, neither are you. How the Monodim manage to move back and forth on their line without bumping into each other and everything else is not yet explained. Somehow they do it, and their versions of light and field strength intensities fall off according to the inverse square law. Don't quibble.

I have chosen the size of 66 billion light-years as a ballpark figure. Actually, it is no more than a wild guess, but I need to have some definite number. One more suitable may turn out to be quite different.

I need to use consistent units in my calculations and chose the cgs (centimeter, gram mass, second) system for that purpose. In the cgs system, the unit of force is the dyne, its dimensions are: $gm \cdot cm/sec^2$. The Relativity constant, c, is $2.998 \cdot 10^{10}$ cm/sec; I often call this light-speed, which is short for the "velocity of propagation of electromagnetic radiation *in vacuo*" and which equals the Relativity constant in number. One light-year is $9.46 \cdot 10^{17}$ cm. You may agree that light-years are handier than centimeters when it comes to visualizing a universe.

GODLINESS: Chapter Four

Now to business: get out your globe. I
will refer many times to a god named Yaul
(rhymes with Paul). You may substitute
your own name for Yaul's once you get the
hang of it. Yaul is a mind and disembodied
head about 28 billion light-years in diameter
happily floating about in empty space. Yaul
is able to "know" where everything is right
now – in defiance of Relativity Theory – to
"see" without light or the delay caused by
light-speed. Yaul can move around, and also
avoid boredom by speeding up, slowing
down or reversing the time stream. To do
such things is just plain magic and Yaul
knows this, and also that an active
imagination is required.

I deal with that big head problem by
looking at an imaginary globe where one
centimeter represents a billion light-years.
Maybe it will work for you.

Yaul creates an imaginary sphere in
space that has North and South Poles 42
billion light-years apart on its axis. It has an
equator 132 billion light-years in
circumference, parallels of latitude and

meridians of longitude. The circle that is the Monodim's universe is placed on Lat 60°N and is moving south at very nearly light-speed. At the center of the Sphere is a Gob of Matter that activates a gravitational field just strong enough to keep each gram of matter/energy in the M'sU (M's for Monodim's, U for universe) orbiting on the sphere's surface between the poles. Got it? Better reread this paragraph a few times.

That Gob of Matter at the center of Yaul's Sphere is a curiosity to which I shall return, but right now I only want to find its mass, M_g. This can be calculated from: the orbital velocity, V_o, which I take as c, or $2.998 \cdot 10^{10}$ cm/sec although it is a trifle less; and the radius of the Sphere, R_s, which is 21 billion light years, equal to $1.99 \cdot 10^{28}$ cm. The formula is:

$$M_g = R_S \cdot V_o / G.$$

Where G is the gravitational constant, $6.67 \cdot 10^{-8}$ dyne·cm^2/gm^2. Solving for M_g, I learn that the central mass is $2.68 \cdot 10^{56}$ gm. That is a mighty big mass – about $1.3 \cdot 10^{23}$ times the mass of our solar system, maybe ninety thousand billion galaxies, more or less.

Yaul numbers the meridians of longitude so that the 0° meridian passes through the homeland of the tribe of Monodim that most interests us. Why not? Ethnocentricity is OK; the English set our Earth's 0° meridian on their observatory at Greenwich, and there it stays.

Yaul turns the time stream backwards 11 billion years and, behold, the Ms'U is squashed up in a tiny space at the North Pole. This is the instant of the Big Bang. From here, all directions are south. Let the time stream run forward 33 billion years and the Ms'U will lie on Lat 0°, the Equator, and will have expanded to its maximum size, 132 billion light-years in circumference.

33 billion years of compression follow, and then the Ms'U is squashed into a tiny space at the South Pole. All directions are north, there's a Big Bang and, in 66 billion more years, it expands and contracts and arrives back at the North Pole. Run through all this again and "see" the Ms'U as a

moving parallel of latitude whose number smoothly changes from 90^oN to 0° to 90^oS to 0° to 90^oN... It does that at the rate of 30 degrees per 11 billion years. Use your godliness to speed up the time flow if you don't want to go nuts with the waiting.

Yaul now puts the Ms'U back where it was: at Lat 60^oN and headed south. It seems to Yaul that the Ms'U is 11 billion light-years away from the last Big Bang. Things are somewhat different for the Monodim. They can only look to the East or the West in their universe, not directly north. Since they are looking back in time as well, it may be better to say that they can effectively look only to the Northeast or the Northwest on the Sphere. That line of sight makes a spiral toward the pole, longer than the direct line by the distance passed through their space. At 11 times the square root of two, they see the Big Bang as 15.6 billion light-years away.

[Insert Animation #1 Here]

We must consider a couple of things that I glossed over. One is that trifle by which the orbital velocity of the Ms'U, v_o, is less than the Relativity constant, c. Since the Ms'U contains matter/energy, its mass, as "seen" by Yaul, would be infinite if its orbital velocity were equal to c. That clearly won't do. Special Relativity shows that the mass of a bit of matter increases with velocity, becoming significant as it approaches light-speed. The formula is:

$$v_o^2/c^2 = 1 - (1/M_{ratio})^2.$$

Where M_{ratio} is the ratio of the amount of matter in the Gob of Mass plus the amount of matter/energy contained in the Ms'U to that latter amount. I'm not a jealous god, but I am a whimsical god. By mere whimsy I have chosen 1,837.5 as the value of M_{ratio}. Some of you may recognize that number from elsewhere. Never mind; I call it a whim.

So the total mass of the Ms'U is $2.68 \cdot 10^{56}$ gm, a bit more than was the Gob alone, and v_o is reduced from c by $1.48 \cdot 10^{-7}$ times c. That is to say, v_o is 0.999999851 times c. The trifle less speed makes the difference between a reasonable mass and an infinite mass.

The mass of the Ms'U, if it were at rest, is thus found to be $1.46 \cdot 10^{53}$ gm, maybe about fifty thousand million galaxies, more or less. It sounds like enough. If I had a dollar for each such galaxy, people might confuse me with El Magnifico, Warren Buffett, Sage of Omaha. But the M'sU is not at rest.

The total mass of the Ms'U, as Yaul "sees" it, is $2.68 \cdot 10^{56}$ gm, which is a trifle more than that Gob of Mass Yaul had once put at the center of the Sphere. There is no coincidence here. Yaul had to get rid of that absurd Gob of Mass which never was a real part of the Ms'U; it was a stepping stone, chuck it out and set v_0 to do the job. Now the Ms'U can orbit around its own center of matter/energy.

Astronomers have long known that the central point of an actual mass will act like a single point source of all the gravitational effect of that mass on space outside of the mass. So long as they deal only with space outside of that actual mass system, their calculations will work out just fine. It's easy to see that the center of mass of the Ms'U lies somewhere on the axis of Yaul's Sphere. The question is, where on that axis? If it is always centered between the poles, Yaul's Sphere will be near perfect, which would be nice.

Suppose Yaul divides the Monodim's matter/energy into two circles instead of one, and locates the second one at Lat 60°S moving north at v_o. Now there are two equal universes, and each has had its Big Bang at the same instant by Yaul's way of "seeing." One fired off at the South Pole when the other did its thing at the North Pole. Clearly the mass center of this double universe is at the center of the Sphere. Yeah, well the double universe is easy to visualize, but...

I'm reminded of the time when Uncle Bertie got suckered into letting a "cosmological constant" slip into Relativity Theory. He rooted it out pretty fast, but it was embarrassing – like a camel in your tent. His "cosmological constant" was a lot more subtle and insidious than this double universe idea. For one thing, Occam's razor says to forget the dual universe. It's just a contrived complication to avoid taxing my brain. I regard my laziness as a virtue; I don't want to make it into a vice. For another thing, what happens when the two universes collide at the equator? Nothing, where nothing meets nothing or nothing meets something. But spectacular fireworks will erupt where something meets something. The fireworks could knock some matter/energy out of both universes, and the system would eventually wind down. Phooey.

Go back to the single universe. We seldom think about the velocity of propagation of a gravitational field. I assert that it is light-speed. If we could find a star going past at near light-speed and could point a directional gravimeter at it, we would see its gravitational field coming from the same place as its light. By Yaul's brand of instantaneous "seeing," the star would be ahead of both its light and its gravitational field.

Consider what happens when the Ms'U is at the Equator, Lat $0°$, and heading south. Its gravitational field will propagate toward the center of Yaul's Sphere, taking 21 billion years to get there. It will take another 21 billion years to get from there back to the surface of the Sphere. During that time the Ms'U will have moved on by two radians of latitude. A radian is 57.3 degrees, so the Ms'U will be at Lat $65.4°$S and heading north when under the influence of this gravitational field from the center of Yaul's Sphere. In general, the matter/energy center of the Ms'U will always be $2 \cdot 57.3 = 114.6°$ behind the latitude of their universe. It might be neater if it lagged $180°$ or $90°$ behind. We shall see.

If the center of gravity of the single Ms'U bobs up and down along the axis of Yaul's Sphere, then the Sphere could be said to bob with it. This is no great problem. Yaul's head could bob up and down in sync. So Yaul could "see" a stationary Sphere with the needed matter/energy functioning from or near its center. If that Sphere should turn out to be a Spheroid, the concept would still work OK.

* * *

I want to pause here. Quite a lot has been accomplished. I have set up a boundless universe of finite size, containing a finite amount of matter/energy. Dividing the amount of matter/energy by the size gives a kooky sort of "density" of $4.29 \cdot 10^{27}$ gm/cm as Yaul "sees" it moving, or $2.34 \cdot 10^{24}$ gm/cm if it were at rest. These large numbers will be cut in half at maximum expansion and vastly increased at maximum compression. This universe cycles between maximum and minimum size in a stable and definite manner. It exists in two dimensions – one plus time – much as our universe exists in four dimensions – three plus time. The meridian lines of orbital velocity are, of course, the time line vectors. They give the directions of the past and the future.

Perhaps you noticed that I have used the Special Theory of Relativity, but not the

General Theory, so far. That was not correct, but it shows how my thinking started out. The major remaining problem is how to transpose from the Monodim's universe to something more like our universe, and back again. Along the way, I must get rid of Euclidean lines and points. They lead to infinite density, which is unacceptable. I now wish to consider the effect of strong gravitational fields, and that calls for General Relativity.

The Monodim's universe, continued

Turn the Monodim's circle into a torus and those pesky Euclidean lines, which become even more pesky Euclidean points at the North and South Poles, just disappear. If I simply say the diameter of a segment of the torus is a suitable number of light-years, the density of their universe could be about the same as ours. The Monodim would then live in a three-dimensional world, and everything would be loverly.

Whoops! Sorry, it's not quite that simple. The M'sU held together mostly because I said it was one-dimensional, in the form of a circle with all its bits orbiting at the same high speed, v_o, along the meridians of Yaul's Sphere. A one-dimensional being can't see any direction or method to get out of that universe. If it could do so, the energy transfer needed to alter its v_o would be very large. Things must be tightened up in a similar way if I'm going to use a torus.

* * *

I played around with the math of the Black Hole In Space concept in the late seventies and later programmed a small Commodore computer to crunch my numbers. The big kahuna is Phi (ϕ): the gravitational field potential. Phi is defined as the work required to move a unit mass from an infinite distance to a point r distant from the effective center of a large mass, M, in an otherwise empty space. Hence:

$$Phi = -G \cdot M/r.$$

Where G is the gravitational constant. This holds true under both Newton's Laws and General Relativity. Also under General Relativity, but not Newton's Laws, Phi is defined as the work required to move a unit mass from the center of a rotating disk to a point r distant from the center; this works out to be:

$$Phi = -\omega^2 \cdot r^2/2.$$

Where ω (omega) is the angular velocity of rotation in radians/sec. By the definitions, Phi can vary between zero and some negative number, so it is most often used as "minus Phi."

A Black Hole In Space (BHIS) is a region in space where a very high –Phi captures any matter/energy that happens to impinge, even tangentially, upon its boundary. The release of any captured matter or energy cannot take place. It is common to consider that such a region is in otherwise free space; and also, that it has a boundary shape (surface of constant Phi) that is spherical. The value of Phi does not affect the radius, r, but it does increase the area of the spherical surface, which is perpendicular to the direction of Phi. Such a shape may be spherical, but not a Euclidean sphere.

* * *

Whew! I just don't know how to avoid this jargon. What's more, I'm going switch units systems on you. A units system defines three units and derives all others from those three. The usual three are distance, mass or force, and time. Distance could be furlongs or cubits, but I will use meters, mtr; mass could be slugs, but I will use kilograms, kg; time is almost always in seconds, sec, as I will take it. This "mks" deal is called the International System. It may be more common nowadays than the cgs system I learned more than half a century ago.

One place where I will ease up is that of formulas and calculations. This is not very professional, but I'm no longer that kind of engineer. I'll try to give more results than number crunches, and hope that mistakes don't slip by. Even reading my old notes makes me bone weary. It must be worse for you. I'll stick most of my math in Appendix A. Some symbols and values are:

BHIS = Black Hole in Space.

c = Relativity Constant, $2.998 \cdot 10^8$ mtr/sec.

Phi = Gravitational Field Potential, in joule/kg.

G = Gravitational Constant, $6.67 \cdot 10^{-11}$ joule·mtr/kg^2.

G·M = A Source of Excitation of Phi, in joule·mtr/kg.

joule = Unit of Work or Energy, in kg·mtr^2/sec^2.

M = Large Mass, in kilograms, kg.

π = Euclidean Circumference/Diameter, 3.14159...

r = Distance, in mtr, from center of G·M.

ω = Angular Velocity, in radian/sec.

A light-year is a handy value for Large Distances; it is $9.46 \cdot 10^{15}$ mtr.

The defining boundary of a BHIS is called an event horizon. Two possible extremes of event horizons are suggested from the definition given for a non-rotating BHIS. One lies where neither matter nor energy can rise above this horizon. At this surface, $-Phi = 0.64872 \cdot c^2 = 5.83 \cdot 10^{16}$ joules/kg and the surface area of the BHIS is infinite. There's that word that I don't trust. The inside of such a Black Hole would be hollow and its $-Phi$ would be nil! How odd. So I take this to be the limiting maximum value of $-Phi$ in much the same sense that c is the limiting maximum value of any velocity.

The term "escape velocity" was popularized after the sputnik furor of the fifties to refer to the least speed a rocket must attain to completely escape our Earth's gravity. The other extreme BHIS event horizon lies where $-Phi$ is just barely large enough to prevent anything from reaching escape velocity. At this surface $-Phi = 0.35390 \cdot c^2 = 3.181 \cdot 10^{16}$ joules/kg and the surface area is finite. Radiant energy (light, x-rays, etc) could rise above such a surface, but would have to fall back sometime. It could not truly escape. This –Phi I take to be the minimum practical value for a BHIS event horizon. At this value, the circumference of a great circle is 1.48337 times what would be determined from r by Euclidean geometry, and the Euclidean surface area would be multiplied by that number squared: 2.2004. The "lag" between the latitude of an orbiting ring of matter and the apparent mass center of such an object would be 77.3°.

So the –Phi of a BHIS can be in the range of $0.35390...c^2$ $(3.181 \cdot 10^{16})$ to $0.64872...c^2$ $(5.831 \cdot 10^{16})$ joules/kg. Since I mean to build such a –Phi along with the torus to come, it is not necessary that the surface of Yaul's Sphere be an actual BHIS.

"In quadrature" is a term that pops up in the design of bass-reflex loudspeakers, electrical engineering, astronomy and elsewhere. It refers to a 90° offset. Yaul's Sphere can remain neatly spherical if I reduce its –Phi to a value where that "lag" is in quadrature, exactly 90°. Such a –Phi = $0.24856...c^2 = 2.234 \cdot 10^{16}$ joules/kg – very large, but not quite large enough to make a true BHIS. At this value, the circumference of a great circle is 1.27324 times what would be determined from R_S by Euclidean geometry.

* * *

The last chapter set up a central mass of $2.68 \cdot 10^{53}$ kg in Yaul's Sphere, acting over an R_S of $1.99 \cdot 10^{26}$ mtr. This gives a value for $-Phi$ of $8.98 \cdot 10^{16}$ joules/kg – impossibly large. So I shall revise my earlier values for the M'sU in order to give them the same or similar properties under General Relativity.

Change the position of the M'sU to Lat 45°N, still moving south. Set the Sphere's radius, R_S, to $1.2241 \cdot 10^{26}$ mtr (12.9 billion light-years). Its exciting mass, M, becomes $4.10 \cdot 10^{52}$ kg. That is 2.61 times the new at-rest mass of the M'sU – enough to keep things neat, although my whimsy of 1,837.5 is shot down.

From Yaul's viewpoint, outside of the Sphere and where the effect of $-Phi$ is negligible, the circumference of a great circle appears to be $2 \cdot \pi \cdot 1.224 \cdot 10^{26}$ =

$7.69 \cdot 10^{26}$ mtr (81.3 billion light-years), and the M'sU appears to be 10.2 billion light-years from the North Pole. But Yaul understands Relativity. On the Sphere that Yaul "sees" after allowing for Relativity effects, the circumference of a great circle becomes $2 \cdot \pi \cdot 1.224 \cdot 10^{26} \cdot 1.27324 = 9.79 \cdot 10^{26}$ mtr (103.5 billion light-years).

And from the Monodim's viewpoint – ignoring any locally induced –Phi – they are 12.9 billion light-years from the North Pole that they cannot see directly, or 18.3 billion light-years from it in the direction they can see. This is because their metersticks are shorter than they would be at Yaul's position, due to their overall high –Phi. If they could "see" it all at once, their universe would be $6.92 \cdot 10^{26}$ mtr in circumference. The multiplier for their stick lengths (and clock rates) is 1/1.27324 or 78.54% relative to a place where total –Phi is trivial.

Now, at last, I'm ready to have a go at the torus – and a tough one it will be. My general idea is that the line circle, which had contained all the matter/energy of a mono-dimensional universe, becomes the locus of the local mass center of the matter/energy scattered around it. It's the mass centerline. My dictionary defines a torus as, "a doughnut-shaped surface generated by a circle rotated about an axis in its plane that does not intersect the circle." I'm using the word "torus" pretty loosely. It's really going to be a toroid, generated by a strange shape. The surface is one of a defined –Phi surrounding the line circle that once was the M'sU, and now is its centerline of mass. It is set up by gravitational fields from both the nearby mass and the more distant, but larger, exciting mass center of Yaul's Sphere. By the way, that mass does not excite Uncle Bertie, you or me; it excites a gravitational field.

* * *

Yaul's magic has been done outside of the Sphere. From a place of negligible –Phi, Yaul magically "sees" without light and magically "sees" everything at once: instantaneously. Yaul would "see" the toroid as if it were generated by a combo-ellipse not a circle. This would be a half-ellipse with its long axis pointed away from the center of the Sphere joined to another half-ellipse with its short axis pointed toward the center of the Sphere. [I'll keep things simple for now and not discuss shape shifts due to the Sphere's curvature. The Monodim would not be aware of the shortening of rods in certain orientations, nor of the other oddities noticed by an outside observer – especially by a god like Yaul.]

The Toroid I am building has its surface boundary at a place where –Phi lies in a range from more than $3.18 \cdot 10^{16}$ joules/kg to less than $5.83 \cdot 10^{16}$ joules/kg. This Toroid is itself a BHIS. Internal matter/energy may move about and be spread out within it. In the nature of things, the Toroid's apex will rollover forward one complete turn as the whole Toroid goes through a full cycle of 131.8 billion years from N Pole to S Pole and back to N Pole. Conservation of momentum leads one to expect that matter near the nadir also will move forward relative to the general motion. This could lead to a backward rotation of internal matter opposed to the forward rotation of shape. I have not yet made these complex calculations; but surface –Phi needs to be more than minimal so that any rotational velocity, plus random movements, cannot boost the speed of surface energy above escape velocity.

The center of mass of this Toroid is a line circle lying on a parallel of latitude of Yaul's Sphere. In other words, it is the once mono-dimensional universe of the Monodim that I have been nattering about. But now the Monodim are free to move around in the three-dimensional space within their Toroid.

Note that the Toroid morphs into an ovoid near the North and South Poles during the time when a Big Crunch becomes a Big Bang. All the matter/energy of the M'sU, $4.10 \cdot 10^{52}$ kg, is then near the Pole. The intersection of this ovoid with the surface of Yaul's Sphere forms a circle around the Pole with a radius of 5.32 billion light-years. This circle lies on Lat $71.5°$. The size of this slightly curved radius depends on the –Phi chosen to define the ovoid's surface. I have taken that to be $5.43 \cdot 10^{16}$ joule/kg at and below the surface of Yaul's Sphere. Below the surface, the ovoid extends 3.42 billion light-years. At the Pole and above the surface of the Sphere, –Phi is also taken as $5.43 \cdot 10^{16}$ joule/kg at the apex of the ovoid, which is 7.23 billion light-years above the Sphere. So the gross size of the ovoid is 10.64 by 10.65 billion light-years. Its shape has been distorted by the addition and subtraction of the Sphere's –Phi along the axis of Yaul's Sphere. Its average diameter is 10.6 billion light-years.

I must now strain myself with a tedious calculation of the volume of this ovoid. The volume is more than that of a Euclidean box that could contain it. Such a box would have a volume of $(10.08 \cdot 10^{25})^2 \cdot 10.07 \cdot 10^{25} = 1.023 \cdot 10^{78}$ mtr^3.

To make volume calculations easier, I set up spreadsheets to figure the General Relativity volumes of a sphere (at Lat 90°N) and a torus (at various latitudes) using an average radius as determined from the values in Appendix A. I assumed that matter was evenly distributed throughout these volumes. The ovoid at Lat 90° comes close to a sphere with a GR volume of $1.92 \cdot 10^{78}$ mtr^3 having an average density of $2.14 \cdot 10^{-26}$ kg/mtr^3.

At Lat 0°, the Equator, the center of excitation of –Phi is not a point but a line. Here the Toroid would rise 2.9 billion light-years above the surface of Yaul's Sphere, at which apex –Phi would be $3.35 \cdot 10^{16}$ joule/kg. It would extend 455 million light-years below that surface. Its half-width at the Sphere's surface would be 774 million light-years. The GR volume of a torus with this average radius would be $1.03 \cdot 10^{78}$ mtr^3 and its density would be $4.0 \cdot 10^{-26}$ kg/mtr^3. This is nearly twice the density at the Pole! Oh dear. I had been expecting the opposite effect. But in the next chapter I show that the Monodim will see their universe as a sphere, not a tube, with a radius that extends back to the Big Bang.

Let's get to the Toroid at Lat 45^oN. Here its apex is 4.68 billion light-years above the surface of Yaul's Sphere; –Phi is $4.08 \cdot 10^{16}$ joule/kg. The Toroid's width is 5.59 billion light-years. It extends 1.71 billion light-years below the surface of the Sphere. The circumference of its circle is that of the parallel of Lat 45°, which equals 73.2 billion light-years. The GR volume of a torus with this average radius is $3.02 \cdot 10^{78}$ mtr^3 – nearly 58% greater than the volume of the polar Ovoid.

This math is not rigorous. The –Phi values are estimates and approximations. These values are highest near the Pole and decrease as the effective mass, M_E, decreases toward the equator. It is just guesswork to relate –Phi to M_E. I chose $3.35 \cdot 10^{16}$ joule/kg for the apex –Phi at the Equator because about 5% greater than the minimal BHIS value felt right. I calculated the height of the apex from that –Phi and the angular velocity; local mass came from the height. Near the Pole, I was content to use a –Phi of $5.43 \cdot 10^{16}$ joule/kg, which came partly from the height of the ovoid's apex at the Pole with the angular velocity and partly from its M_E, which equaled M_U. At other latitudes, I found the apex's height and –Phi from the geometry of an ellipse. The local masses sort of fell into place. My math can be found in Appendix A. All this involved a number of assumptions – or presumptions.

* * *

It turns out that I cannot reconcile the idea of very high density at the time of a Big Bang with a universe that does not contain an infinite amount of matter/energy. Is this important? Perhaps not.

As the Monodim's toroidal universe approaches the North Pole, it starts to collide with itself past Lat 72°. Actual head-on collisions of matter release an enormous amount of energy. They approach the science fiction idea of a matter to antimatter collision. Such collisions will be relatively rare at first, but their energy release will tend to gasify nearby matter and increase the local collision rate. The general collision rate will increase for near five billion years. At that time, the entire universe surrounds the Pole. As this Ovoid continues to morph back into a Toroid, the collision rate will decrease until it stops completely before Lat 72°. The whole collision period takes near ten billion years. Such a long time at high temperature and low density may or may not have the same effect as Gamow's short time at higher temperature and high density. So far, I'm betting that it will.

[Insert Animation #2 Here]

There is more to be done. But let's not get bogged down in detail before looking at this universe from the Monodim's viewpoint. We do so in the next chapter.

Let's move Yaul from that rather aloof position outside of the Sphere and the Monodim's universe – outside and looking in, so to speak – to inside and looking around. Yaul then becomes a part of that universe although still a god. We can also change Yaul's size and shape to that of a typical Monodim. Let's change the name to "I" or "we" or "you" as the grammar warrants. It seems more personal that way.

* * *

I find myself on a planet that looks and feels just like Earth. I am among a lot of Monodim who say they are Americans. They speak a language they call English, although I come to learn that another tribe, the British, dispute their claim. They accept me as one of themselves. Why all this should be is a mystery, but I'm glad of it. It becomes easy for me to use their libraries and observatories. Of course, my godly powers of observation may be even better than their telescopes. Right away I note that this planet orbits around a star they call Sol and has a moon of its own. There are other planets in this solar system. Also there are other stars, some with planetary systems, farther away and scattered about. There are star clusters as well. All these objects are loosely arranged in a kind of big system called a galaxy. There are other galaxies even more distant, but this local one is named the Milky Way.

As I notice details around me, I see that metersticks have the same length whatever their orientation. Circles and wheels remain circular when placed at every possible angle. This seems odd until I remember that I am now an inside observer, not an outside one. My watch runs normally and measures my pulse rate the same as always – and for the same reason. There do not appear to be any abnormalities. I ask after Uncle Bertie, but nobody seems to know him by that name. Many have heard good things of Albert, who has died much lamented; some of the more scientific types are familiar with his General Relativity Theory.

Next, my attention turns to astronomy. The night sky is full of stars and constellations in their same old comfortable positions. But wait – I'm looking with the naked eye, so only super bright or fairly nearby stars are visible. I then use my godly abilities to find my position on Yaul's Sphere. It turns out that I am 100 million light-years directly above the surface point at Lat 45.12°N, Long 0.00°. This means that I am about 1.8 billion light-years away from the nearest surface of the Toroid, a surface where –Phi is $4.1 \cdot 10^{16}$ joule/kg. I'll need a big and powerful telescope to get out past that distance. While making arrangements to borrow such an instrument, I'm going to digress with a little lecture on vision.

Until a very few centuries ago, people thought – if they thought about it at all – that sight or vision was the result of some sort of rays that emanated from the eyes. This gave rise to a number of weird beliefs such as: dread of the "evil eye" and the notion that "how far can you see?" was a valid question. One of Baron Munchausen's pals saw a guy sleeping under a tree several hundred miles away. Even now, Superman has "X-ray" vision and the ability to project "laser beams" from his eyes.

Today we know that sight is our perception of light emitted from or reflected by objects. High schools teach that our eyes are similar to cameras. They often fail to mention that these are very crude cameras, with lumpy lenses and distorted focal planes, quite incapable of snapping the clear pictures we see. In fact, our vision is the result of the mind (brain) processing data that streams in from the eyes' moving sensors, plus short- and long-term memory and knowledge.

We never actually look through powerful telescopes. Instead we take pictures, often using a wider spectrum than visible light, including X-rays. We also take readings from spectrometers. We use radio telescopes for low frequency stuff. We try to make sense of all these data. So do the Monodim. But, for the moment, I will revert to that old idea of sending out a ray. I won't project it from my eyes. No, I will squirt it out of an immensely powerful and hypothetically perfect laser pointer that I happen to have in my pocket. First, I will locate a plane that contains and is determined by the meridian of Long 0° on Yaul's Sphere.

I aim the laser pointer within this plane at the nearest perpendicular surface of the Toroid. This sounds like a dangerous and foolhardy experiment. That laser beam is going to slow down, exit the Toroid, stop and then come back. It could hit me smack in my aiming eye. But it will take over four billion years to do this. During such a long time, I might blink or wander off somewhere or I might even die. Also, that aiming plane is far from perfect, neither is it constant. The gravitational fields from the Toroid's lumpy contents distort it. Those lumps move about with their own velocities of translation. So the aiming plane's imperfections are in continual flux. Furthermore, there is no such thing as a hypothetically perfect laser. I'm not worried.

This reminds me of the lawyer who was not worried about losing his case. His argument went along these lines, "Your

Honor, my client's dog does not bite people. Also, my client's dog was out of town at the time the alleged incident took place. Furthermore, my client does not own a dog."

A gravitational field deflects a ray of light – or any other electromagnetic radiation – in much the same way that the refractive quality of glass deflects a ray of light that passes through a glass lens. As it happens, this effect was found long ago in the first predicted experiment to confirm the General Theory of Relativity.

If I aim my laser pointer ever so slightly away from the "perfect" aiming plane, there will be an East or West component in its path. That east or west drift will increase as the laser ray passes through or beyond the Toroid. The overall effect is that: whatever direction I may aim my laser pointer, its beam comes back to the Toroid, if it ever left, and then loops along to the East or West within the Toroid. The only angles where this would not be true are so narrow as to be non-existent.

Inside the Monodim's universe

Enough with the laser pointer! Let's go back toward reality and observe the points of light that are stars and galaxies. Our observations will be done the same way as they would here in our world – with optical and radio telescopes.

Start by looking in the general direction of a band centered on that old aiming plane. We will look out as far as the Toroid's surface. This distance ranges from more than one-and-a-half to more than four-and-a-half billion light-years. It's quite a bit more, due both to the slowing of clock rates at high –Phi and whatever angle there may be, known to us gods, from that aiming plane. Far distant stellar objects can only be seen as point sources of light or radio waves. We might see them become more numerous in the space close to the Toroid's surface. If we looked out past that surface, we would find an empty region and then start to see stellar objects again. But there is a very big problem with this scenario. How can we tell the actual distance to a far object? It isn't easy.

In astronomy, we can quite accurately measure the relative brightness of a stellar object. If we knew its absolute brightness, we could calculate its distance, assuming none of its light had been absorbed in transit. But we do not know its absolute brightness, so we try to estimate the value. There are several tools to help. They include spectral analysis of the incoming light and other radiation, measures of pulsation if any, Hubble's constant for the expanding universe, theories of stellar evolution over time, assumptions about the shape of space, and statistical math to tie it all together. Some of the reasoning is circular. Like I said, "It ain't easy... and the number could be pretty far off, even if in the ballpark."

Suppose we point our telescope to the west along the tube, so to speak, instead of at the Toroid's surface. Because we are looking back in time, we are looking northwest toward the remnants of that Big Slow Bang that may appear to have happened sometime between 14 and 18 billion years ago. Along the way, we see stellar objects in what passes for a near straight line in this universe.

Let's swing the angle 40° and take a peek at a galaxy named Arthur, 3 billion light-years away. Although Arthur is a fair sized galaxy, she is so far away that she appears only as a point of light – no structure can be discerned. Arthur shows the expected "red shift" and has the peculiarity of emitting a pulse of radio waves every so often. It is fairly easy to identify Arthur because of this pulse frequency.

With the help of a god highly skilled in telescope pointing, we see another picture of Arthur from a quite different angle, but still to the west of that old aiming plane. This version of Arthur is 9 billion light-years away, showing when he was very young. [Through a sudden flashback of omniscience, I remember that galaxies are males until puberty at which time they change sex to female. This is disconcerting to them, and may have something to do with growing spiral arms and a nasty disposition.]

A third view, at yet another angle, shows Arthur to be 6 billion light-years distant.

So the same galaxy might appear at three different places in the night sky. Sadly, it is next to impossible to determine that these three views are of the same galaxy. Few, if any, galaxies actually pulsate. If one did pulsate, it could hardly be expected to keep up the same identifying rate over a time span of 6 billion years.

Whatever direction we point our telescope, we end up looking back in time at objects within the Toroid to the northeast or northwest of our position. We can look back as far as the Big Long Bang if our telescope is big enough. It appears to us that we are looking from the center of a sphere whose radius extends all the way back to the Big Multi-Bang. The perceived volume of this universe can be determined from that radius.

The problems are multiple. Although I located our telescope near the center of the Toroid and first suggested that we look straight west, I don't know when and where to pin down the point of a Big Bu-Bu-Bu-Boom that went on for ten billion years.

Other directions of observation give a sight line that swings back and forth across the Toroid and passes through regions of high –Phi, slightly confusing that "universal radius." If we were observing from a place closer to the boundary of the Toroid, most all of our sight lines would swing back and forth. So it goes.

But the cross-section of the Toroid is relatively large and we can't get much more than three crossings, if that many, before arriving at the Big Long Bang. Similarly, we are limited in the number of multiple sightings of the same object. I can roughly estimate the perceived volume and density of the M'sU by taking 15 billion light-years as the radius, Yaul's known Mass of the Universe as $4.1 \cdot 10^{52}$ kg and a simple average of Yaul's calculated –Phi $= 4.8 \cdot 10^{16}$ joule/kg. Plugging these numbers into my General Relativity spreadsheet gives: Volume $= 2.9 \cdot 10^{79}$ mtr^3 and Density $= 1.4 \cdot 10^{-27}$ kg/mtr^3. That apparent volume is nearly ten times the General Relativity volume calculated for the Toroid at Lat 45°.

* * *

There really is a great deal more work to be done. Among other things, I should set up a computer animation to show how light rays travel between objects in the Toroid. The trouble is that word: "work." I've become slow and unfocused. It's too tedious, and I'm too lazy. Maybe later. Yet I may have gone far enough to paraphrase Pogo, the savvy possum from that long gone eponymous comic strip, and say, "We have met the Monodim and they is us."

T his chapter got its name, *Loose ends*, long before I wrote it. Such a subject could fill the Library of Congress and run out into the street. But the only loose ends I want to talk about are those to do with my cosmology. My remarks on godliness are nearly complete.

I have built a universe that, from within, appears very much the same as our universe did seventy-five years ago. It has seven remarkable qualities. These are:

1) The total amount of its matter/energy is finite.

2) Every location is moving at a very high velocity, $0.7854 \cdot c$, on a collision course with its counterpart, which is exactly $180°$ (of longitude) around the universe.

3) Because of the very high velocity, over 23% of the mass of any and all matter is due to its kinetic energy.

4) Every location collides once with its exact counterpart in a definite cyclical period of time.

5) The overall collision period of all locations (the Big Bang period) is about one-eighth of the cyclical period noted in (4).

6) The apparent radius of this universe is the distance to the Big Bang period center, and its apparent volume is that of a sphere of this radius.

7) Due to multiple sight angles for the same galaxy at great and greater distances, density will appear to increase faster with distance from the observer than predicted by simple geometry.

Further detail and reworking may bring this model into accord with current observations.

Many people believe that natural laws as described by Newton, Galileo and Kepler are higher on the pecking order of things than theories like Uncle Bertie's Theory of General Relativity. Not so. Those older "laws" date back to a time when reality was thought to be straightforward and definite, governed by immutable law. By the advent of General Relativity, we knew there were wheels within wheels. Uncle Bertie devised a hypothesis that became a theory when it was shown to embrace the known facts. It became a Theory with a capital T when it predicted unknown facts that were later confirmed. It is higher on the pecking order than Newton's laws – which it embraces.

This present cosmology is a proposed geometry of the universe, not quite a complete hypothesis, far less a theory. It still lacks a computer work-up that will clearly show the night sky from the viewpoint of an astronomer at various locations within the universe, and some other things. Note that I cannot estimate apparent density without that computer work. But Hubble's constant should come out about right.

My geometry of the Monodim's Universe filled the last three chapters, and was itself filled with a lot of loose ends and dangling what-ifs. This came about partly because, for decades, I lived on a small Caribbean island where it was not easy to get up-to-date scientific information. It was possible, just not easy – so I didn't do it.

GODLINESS: Chapter Seven

Now, I am close to many Universities and Libraries and even have a high-speed line to bring the Internet into my home. Yet I've scrupulously avoided seeking out what is currently accepted in cosmology. The reason is simple: I want this essay to represent only my own thinking, based on Gamow's book and Einstein's writings.

There are a few other sources. Popular stories meant for a general audience gave me the basic definition of a Black Hole in Space, but no more. There also were ongoing estimates of the time since the Big Bang. A millennium ago, the age of the Judeo-Christian World could be found from Scripture to be headed for five thousand years. A century ago, the age of the Earth, and maybe the universe, was set at 150 million years by paleontology, not astronomy. Half a century ago, Gamow's book put the time since the Big Bang at three and one-third billion years. Over the decades since, that estimate gradually rose to a peak of around 25 billion years and then dropped to about 15 billion years, last I heard.

It seems likely that others would have come up with thoughts similar or identical to mine, but I have not heard of them. Also, I have never heard of a torus orbiting anything, much less itself. To build a geometry based on Relativity was more popular in the mid-twentieth century than now. I'm tooting an old horn. To work alone, without anyone to crosscheck your notions, is asking for trouble. I may have strayed into some error invisible to me, but glaring to another's view. So my ideas could hold a deadly blunder that I have missed. I've never seen any detailed info about Black Holes, although plenty should exist. Or maybe people have bogged down on the subject.

I sense a reluctance to speculate on conditions inside of a BHIS. It appears that small Black Holes do not exist. A large one may be located at the center of the Milky Way and other galaxies. It certainly would be handy to know what is the minimum mass for any Black Hole and why. I've described an entire universe located inside an oddly shaped, and shape shifting, Black Hole. If smaller Black Holes occur within this universe, what happens to them as my Toroid goes through its convolutions? Do they accumulate? If so, the cyclical rebirth might be only partial – a bit contrary to my ideas.

The Hiroshima atom bomb converted about one and one-half grams of matter into energy. A head on collision between two large Black Holes near the Pole could convert more than 23% of their mass into energy. Such a fantastic explosion might not let matter/energy escape from the new Black Hole, but it certainly could vastly distort its boundary shape. It could stretch out to the point of breaking up into numerous smaller Black Holes. In other words: it could splatter. The smaller Black Holes might become the nuclei of new galaxies.

In Chapter 4, I rather cavalierly dismissed the idea of two equal sub-universes – one setting off from the South

Pole at the "same instant" that the other leaves the North Pole. I said their collision at the Equator would scatter matter/energy outside of both of them. This objection is not valid when each sub-universe is contained in a toroidal Black Hole. Their mutual collision at the Equator might last about 650 million years while each one's self-collision at a Pole could take near ten times longer. The former might be called a Pretty Big Bang while the latter would be the Bigger Bang. From Lat 45°S moving south we would look back toward a Pretty Big Bang and see more multiple images of the same galaxies than we would from Lat 45°N moving south. This would tend to increase estimates of density at a large distance.

If one considers the single universe model as somewhat analogous to a hydrogen atom, then the dual universe model would be somewhat analogous to a helium atom. It certainly is easier to visualize the center of gravity of a dual universe as a point fixed at the center of Yaul's Sphere. In either case, the "in quadrature" concept is needed to put a strong gravity vector at near right angles to the gross movement of the Toroid. Although it might make sense to work through the dual universe model, I save myself a bit of trouble by sticking with the single universe idea – at least for the present.

While trying to calculate the shape and volume of my universe, I made a bunch of simplifying assumptions. Perhaps the rawest of these was to reduce –Phi from $5.43 \cdot 10^{16}$ joule/kg at the Pole to $3.35 \cdot 10^{16}$ joule/kg at the Equator. The former value may be partly justified by equation (7) in Appendix A. The latter value is wholly arbitrary. I had the vague thought that the event horizon surface of a Black Hole might show a –Phi partly dependent on the mass of that Black Hole. That is to say, some minimum mass might give rise to a minimal –Phi, and an increase of this mass might tend to increase –Phi as well as the radius of the Black Hole. I have no real justification.

Another, maybe better, vague thought is that the surface area of my shape shifting toroid should remain constant – or is it the volume? At Latitudes $0°$, $45°$ and $90°$, the areas are in ratio of 2.16 : 4.10 : 4.64 ($\cdot 10^{53}$ mtr^2); the volumes are in ratio of 1.03 : 3.02 : 1.92 ($\cdot 10^{78}$ mtr^3). The constant volume idea does not look promising, but other values of –Phi and M_E at $0°$ and $45°$ could bring those area ratios together. I have not done it. Why should I? It's a valid question, and I'm tempted to go back and modify for constant area. Yet such modifications disguise my original thinking. They may be better left for the time of writing an optical simulation, but I'm giving them a go in Chapter Eight.

To develop, at this early stage, a rigid sort of toroidal shape linked to –Phi could lock up my universe with a single possible total mass. That sounds neat, but what if it didn't jibe with observed facts? I would rather trim theory to fit the observations – another good reason for delay.

I expect that quite a few loose ends will be ferreted out and resolved in the course of setting up algorithms for a more complete simulation. Running the simulation on a "what if" basis should give added insights. For example: "What if we double this number, or halve that one?"

There are two numbers that define the size of our universe and its position on Yaul's Sphere. These values were arbitrarily assigned as: Mass of Universe equals $4.10 \cdot 10^{52}$ kilograms and Time from Last Big Bang equals about 15 billion years. The "in quadrature" assumption requires that $-$Phi be equal to $2.234 \cdot 10^{16}$ joule/kg at the surface of Yaul's Sphere. This defines the velocity of the Toroid's orbit. Together with the chosen Mass, it defines the radius of the Sphere. The Time from Last Big Bang then gives the Toroid's current Latitude. In the doing, I set Latitude at 45°N and let the Time from Last Big Bang work itself out to about 18+ billion years.

You can see that this system is much constrained. Of course there is some wiggle room. The Sphere may be less than perfect. The 90° in quadrature may be a bit plus or minus. It may not be feasible to observe the exact time since the last Big Bang. And the latitude can best be known by matching a simulation to our known skies.

A child of hope, I have assumed that matter/energy is distributed more or less uniformly throughout the volume of the Toroid. The units of matter are not atoms or ions, stars or planets, but more like galaxies. They will have their own random velocities of translation, but their average movement is that of the whole Toroid, including the rollover of its matter. While the mass centerline of the Toroid lies on the surface of Yaul's Sphere, the shape centerline is above the surface. This gives rise to a density difference above and below that surface. Conservation of momentum has the upper matter moving back as the lower matter moves forward, causing the rollover.

This rollover will give rise to an effect analogous to the Coriolis force on Earth. While the Toroid moves southward: an object moving to the West will appear to spiral counterclockwise; moving to the East, it will appear to spiral clockwise. The spirals are reversed when the Toroid is moving northward.

This version of the Coriolis force could have an effect on spiral galaxies, but so could other, more random, forces. My excursion away from infinite matter has turned into a mighty complex form.

Disk-like or spiral galaxies near the boundaries of the Toroid would appear elliptical unless at right angles to a radial sight line. Statistics might show what seems an odd distribution of viewed angles of such galaxies. This sort of thing could be a powerful tool for refining my geometry.

*

I recently read something about very rare night sky explosions much, much bigger than any supernova. If these are old enough, they might be head on collisions of major objects near Lat 75° where the final change from Polar Ovoid to Toroid takes place.

T he Monodim seem to have taken to heart some of my remarks on the layout of their universe. They recited a little haiku for me. It goes like this:

Our universe is

A furling doughnut that was

A bright seething sphere.

I am touched by this simple tribute and inspired to get busy and improve my math on the Toroid, its rollover and size.

The algebraic values and conventions used in Appendix A are repeated here:

BHIS = Black Hole in Space.
c = Relativity Constant, 2.998×10^8 mtr/sec.
E = Energy or Work, in joules.
e = Base of Natural Logarithms, 2.71828...
f = Force, in newtons = N = kg·mtr/sec^2.
ϕ = Gravitational Field Potential, in joule/kg.
G = Gravitational Constant, 6.67×10^{-11} joule·mtr/kg^2.
GM = A Source of Excitation of Phi, in joule·mtr/kg.
joule = Unit of Work or Energy, N·mtr = kg·mtr^2/sec^2.
light-year = 9.4611×10^{15} mtr.
M = Large Mass, in kilograms = kg.
m = Small Mass, in kilograms = kg.
π = Euclidean Circumference/Diameter, 3.14159...
R, r = Distance, in mtr, from center of GM.
ω = Angular Velocity, in radian/sec.

* * *

GODLINESS: Chapter Eight

Let's start with a few assumptions that are meant to simplify things, but which may be altered as we go along. These are:

1) We observe ("see") all parts of this universe simultaneously, without using light. In the first instance, we "see" it from the outside where $-\phi$ is zero and the surface of Yaul's Sphere shows a $-\phi$ of $2.234 \cdot 10^{16}$ joule/kg. In the second instance, we "see" it from its orbiting mass center on the surface of Yaul's Sphere where $-\phi$ appears as zero.

2) The Toroid's shape is a torus of circular cross-section. Similarly, the polar Ovoid is a sphere.

3) At constant $-\phi$, the surface area of the Toroid remains constant as it goes through its whole cycle.

4) The Toroid's rollover period at the Pole is equal to its whole cycle period. The rollover period alters itself at other latitudes so as to maintain constant velocity.

Polishing the Toroid

For the moment, I also assume that the Toroid's boundary $-\phi$ is $5.43 \cdot 10^{16}$ joule/kg as found from equation (7) in Appendix A; M_U is $4.10 \cdot 10^{52}$ kg and R_S is $1.224 \cdot 10^{26}$ mtr as in Appendix A. From this we know that the radius, r_{lat}, of the polar Ovoid/sphere at Lat $90°$, r_{90}, is found from $-\phi = \dfrac{GM_U}{r_{90}}$ to be $5.038 \cdot 10^{25}$ mtr. Using Assumption (3), we can calculate the radius of the toroid, r_{lat}, at any latitude from the formulas for area of a torus, $A = (2\pi R_{lat})(2\pi r_{lat})$, and of a sphere, $A = 4\pi r_{90}{}^2$. General Relativity effects will increase these Euclidian areas by a multiplier function of $-\phi$. That multiplier is $1/(1 - \phi/\phi_{max})$ for the sphere. When $-\phi = 5.43 \cdot 10^{16}$ joule/kg, the multiplier = 14.56. It is the same for the $(2\pi r_{lat})$ portion of the torus formula, but works out to be $1/\sqrt{1 - \dfrac{2.2339}{5.8304}} = 1.2732$ for the $(2\pi R_{lat})$ portion.

To ease the tedium of calculation and to aid in cross checks of my math, I set up a spreadsheet. It contains many formulas and is locked except for four cells that allow entries for: single or dual universe, total mass of universe, torus' surface –Phi and torus' latitude. The number cells are coded to show the observer's viewpoint. Numbers good from both viewpoints are in **bold**. Those "seen" distant from the Sphere where –Phi = 0 are in _italic._ Those viewed from the mass centerline, where –Phi again is 0, are in standard type. <u>Underlined</u> numbers indicate mixed or confused viewpoints.

It's too bad that I cannot include the spreadsheet here. It really helps avoid mistakes due to mixed viewpoints and upside down numbers. I do list my more important results. The numbers and formulas include: $c = 299792458$, $G = 6.672 \cdot 10^{-11}$, $-\phi_{\max} = (\sqrt{e} - 1) \cdot c^2$, $-\phi_{quad} = -\phi_{\max} \cdot (16 - \pi^2)/16$; $\omega = (-\phi \cdot 2/r^2)^{0.5} \cdot ((1 + \phi/\phi_{\max}) \cdot (1 - \phi/\phi_{\max})^{1/4.4516})^{0.5}$, $v_o = \omega \cdot r$. The power of 1/4.4516 for the last term in the formula for omega was found by trial and error to adjust ω to the right value for the quadrature –Phi. I spent a lot of time bogged down on this faulty equation. Then I decided to move on, so flagged it red (orange for its offspring) on my spreadsheet. More later. The Newtonian formula for orbiting velocity is much simpler: $v_o = \sqrt{-\phi}$. Compare it to the Newtonian formula for escape velocity: $v_e = \sqrt{-2 \cdot \phi}$. These don't work at high values of –Phi because velocity contains the omega time unit ($v = \omega \cdot r$) and is observed from a confused viewpoint. I already set up a few high –Phi velocities. At just under $-\phi_{max}$, v_o = c. Right at $-\phi_{max}$, v_o becomes meaningless as the surface area of the BHIS becomes infinite. Just on $-\phi_{min}$, v_e = c. Right at $-\phi_{quad}$, v_o = c·(π/4).

As viewed away from and outside of Yaul's Sphere, the shape center of the torus is displaced outward along the radius of the Sphere from its mass center. This is due to the "in quadrature" gravity vector from the center of Yaul's Sphere. This gravity vector creates both a time and a distance multiplier of 1.2732 on the surface. There are other effects. Total displacement, Δ_{lat}, might be calculated. But I will <u>ignore any such displacement</u> for the time being, in accord with Assumption (2); and also because our main interest is within the universe.

Our torus is centered on the surface of Yaul's Sphere where $R_{lat} = 1.2732\cos(lat)(R_S)$ and R_S is the radius of Yaul's Sphere. Volume calculations require knowledge of matter/energy distribution within the volume. <u>I assume that density is uniform.</u>

A sphere is made up from very thin concentric shells. Each shell contributes its volume to the total volume and its mass to the total mass. The total mass inside of any shell creates the $-\phi$ of that shell which increases its volume by the multiplier mentioned: $1/(1-\phi/\phi_{max})$. The cumulative effect of these multipliers is a multiplier that applies to the Euclidian formula for volume of a sphere. From calculus:

$$\int \frac{xdx}{ax+b} = \frac{x}{a} - \frac{b}{a^2}\ln(ax+b) + C. \ \ Let \ x = r, a = -\frac{\phi_r}{\phi_{max}},$$

$$b = 1, \ C = 1. \qquad Then \ \int_{r=0}^{r=1} = \frac{1}{a} - \frac{1}{a^2}\ln(1+a) + 1.$$

ϕ_{max} = $-5.8304 \cdot 10^{16}$; let ϕ_r = $-5.43 \cdot 10^{16}$. Solving for $1/a = -1.074$, $a = -0.931$ and the multiplier is 3.01.

A torus is made up from a series of thin circular disks. Each disk has a Euclidian area of πr_{lat}^2. This area, if multiplied by a center thickness of one unit, is the specific volume of a disk. The same multiplier used above for a sphere thus applies to the Euclidian volume of a disk. Total volume of the torus is then found from the number of such disks: $(2\pi R_{lat})$.

In case you want to compare numbers with Chapter Five or the Appendix, for $-\phi$ = $5.43 \cdot 10^{16}$ joule/kg, we find:
Length/Area multiplier = 3.816/14.561.
Volume multiplier = 3.0142.

r_{90} = $5.04 \cdot 10^{25}$ mtr, V_{90} = $1.61 \cdot 10^{78}$ mtr^3.
r_{45} = $7.33 \cdot 10^{24}$ mtr, V_{45} = $3.52 \cdot 10^{77}$ mtr^3.
r_{00} = $5.18 \cdot 10^{24}$ mtr, V_{00} = $2.49 \cdot 10^{77}$ mtr^3.

As a curiosity, let's calculate r and V for a torus where r = R.
$4\pi(5.036 \cdot 10^{25})^2 = 4\pi^2 r^2$, *so* $r = 2.8413 \cdot 10^{25}\, mtr$.
V = $1.365 \cdot 10^{78}$ mtr^3.

It might be instructive to calculate using $-\phi = 3.35 \cdot 10^{16}$ joule/kg instead of 5.43...:
Length/Area multiplier = 1.5331/2.3506.
Volume multiplier = 1.8484.
$r_{90} = 8.66 \cdot 10^{25}$ mtr, $V_{90} = 4.22 \cdot 10^{78}$ mtr^3.
$r_{00} = 1.36 \cdot 10^{25}$ mtr, $V_{00} = 1.05 \cdot 10^{78}$ mtr^3.

*

Let's take a look at rollover periods – what the Monodim call the furling of the torus. A rollover may help to keep the Toroid's contents from falling in toward the mass center. But it's not strictly necessary. A one kilogram mass at the mass center will have less mass if moved to the surface boundary – see Equation (1) in Appendix A. If that same mass is put into orbit just above the surface boundary, it will gain enough kinetic energy to again become a kilogram. So rollover is a tradeoff as far as distribution of the Toroid's contents is concerned.

Rollover does serve to set things up so that all head-on components of collision velocity are equal at the Pole. Let r_{90} be the radius of the torus (when a sphere) at the

Pole, R_S is the radius of Yaul's Sphere. We need to multiply R_S by 1.2732 to adjust for the Monodim's shorter metersticks. Then the ratio of r_{90} to the enlarged R_S is the rollover velocity fraction. Rollover velocity will stay constant while latitude changes. Rollover frequency will increase as r_{lat} gets smaller.

Rollover amounts to a spinning of the torus around its core, a furling of the BHIS. To avoid loss of energy/matter from this universe, minimal −Phi must increase to at least $4 \cdot 10^{16}$ joule/kg. I no longer see any reason for a −Phi so large as $5.43 \cdot 10^{16}$ joule/kg. When I look back at equation (7) in the Appendix, it is arbitrary. So I now set −Phi to $\underline{4.4678446 \cdot 10^{16}}$ joule/kg. This peculiar number was found by trial and error to give a polar to equatorial radius ratio of eight to one.

Such magic numbers sometimes pay off. I noticed that my faulty equation said that $v_o = 3.3743 \cdot 10^8$. This is wrong. It should be a bit less than c. The fraction of rollover velocity to v_o read 0.4238 and I guessed that it properly would be 0.50/1.2732 (0.3927). From here I back-calculated v_0 as $2.8598 \cdot 10^8$ mtr/sec when –Phi is 4.4678..., a likely sort of number. Effective net –Phi for the BHIS (after allowing for the rollover) now came out to $3.78 \cdot 10^{16}$ joule/kg – plenty over minimal for comfort. A couple of days later I noticed that my trial and error –Phi was exactly twice the –Phi for quadrature. It's great when a plan comes together.

My poor v_0 equation was then reworked:

$$v_o = \left(\frac{c^2}{r^2} \frac{-\phi}{-\phi_{max}} \left(\frac{-\phi_{max}}{c^2} + 0.40744 \right) \Big/ \left(\frac{-\phi}{c^2} + 0.40744 \right) \right)^{0.5} r.$$

It's still arbitrary, and doubles up at low values of –Phi, but it's good enough in the high range; and I use it hardly at all. That's Engineering!

I'm pretty sure that this bit of math could have been done better in minutes or an hour during my twenties. That leads to another general observation: Young engineers may be sharp, but they're erratic. Old gods tend to bumble. So watch out!

*

Polishing the Toroid

While I'm at it, I'll resize the Monodim's universe so that their astronomy will tell them they are about 15 billion light-years removed from their Big Bang. The spreadsheet makes this calculation easy, but there are still things to be decided. Time to the Big Bang is a tradeoff between the total mass of their universe and its Latitude. One defines the other. I first chose 45°N because I wanted to be well away from the Polar Sphere and the Equator. Since the Polar sphere is relatively larger with the new –Phi, I now set current Latitude at **42°N**. – still arbitrary, but reasonable and not so pat as an even eighth of a circle.

42°N Latitude is 48° from the North Pole – 4/15ths of a half circle. The Monodim will see this as 15 billion light-years if the whole circle appears to be about 112 billion light-years in circumference. So I set the total mass of their universe to **2.75·10^{52} kg.** From Yaul's viewpoint outside and away from the Sphere: Radius of the Sphere is *8.21·10^{25}* mtr. The circumference of a great circle is

$6.94 \cdot 10^{10}$ light-years, but the period of time taken to orbit that circle is $8.84 \cdot 10^{10}$ years.

From the Monodim viewpoint, riding on the mass centerline of their torus: The period of time taken to orbit Yaul's Sphere from North Pole to South Pole and back to the North Pole is $1.126 \cdot 10^{11}$ years. They are 15.01 billion years from the Pole, which they see as a 15 billion light-year radius for their universe. Their torus has a minor radius of $6.91 \cdot 10^{24}$ mtr. Its major radius is $7.77 \cdot 10^{25}$ mtr. Its boundary is where $-Phi = 4.468 \cdot 10^{16}$ joule/kg and its area is $9.07 \cdot 10^{52}$ mtr^2. Its volume is $1.59 \cdot 10^{77}$ mtr^3.

The Monodim's $-Phi$ effective for Black Hole conditions is reduced by rollover velocity to a net useful $3.78 \cdot 10^{16}$ joule/kg. The ratio between their Polar radius and Equatorial radius (separated by 28 billion years) is eight to one. The ratio between Polar and Equatorial volumes is five and one-third to one, but area remains a constant $9.07 \cdot 10^{52}$ mtr^2 at all latitudes. Self-collision starts, and stops, at Lat 77°.

*

Polishing the Toroid

Suppose I now switch my spreadsheet over to a dual universe system. Remember that each universe has half the mass of the single system. Each sets forth from its Pole at the same moment, according to Yaul's way of "seeing." They will meet at the Equator, and pass through each other with lots of fireworks. They are identical and I will only talk about one of them.

Monodim astronomy will still show they are 15 billion light-years from a Big Bang. It will not show them that a Smaller Bang is due in 13 billion years. Their apparent density is somewhat more than half what it was in the single universe. That extra is due to more multiple sightings of the same objects along the narrower torus. The ratio between Polar and Equatorial radii has doubled to 16 to 1. Also, the rollover velocity fraction has halved and the net effective $-$Phi has increased to $4.30 \cdot 10^{16}$ joule/kg. It might improve our comparison to set up the same net $-$Phi we found for our single universe.

GODLINESS: Chapter Eight

I switch input −Phi to $3.97 \cdot 10^{16}$ joule/kg and get a net −Phi of $3.78 \cdot 10^{16}$ joule/kg with a velocity fraction of 0.221. The Polar to Equatorial radii ratio is 14.22, and the current volume is $1.46 \cdot 10^{76}$ mtr^3.

Which is more likely: the single or dual universe? Take your pick. I prefer the single, but only through vague intuition. The dual seems simpler, because it has a permanent fixed center of gravity.

* * *

Speaking of likelihood, if the Monodim are located at the mass centerline of their universe, they are at the most special place of all. Everything they can see is at a higher −Phi, up to $4.47 \cdot 10^{16}$ joule/kg. If their location is random, it probably is near the middle volume spot, 87% of the radius away from the centerline. Then their local −Phi is about $2.6 \cdot 10^{16}$ joule/kg. It's likely that the things they observe are at a −Phi from $1.9 \cdot 10^{16}$ above to $2.6 \cdot 10^{16}$ joule/kg below their own value. More complications!

Cosmology may seem a very large subject, but it is a bit abstract – or is it abstruse? I guess it's both. Who the hell cares, anyway?

Well I do, for one. If all, or any part, of my cosmology gains acceptance, it would be good for my ego. This shows how small a subject cosmology can be. Some may argue that my ego does not need a boost. But maybe yours does, and I've been trying to show you how to become a god. I'm also suggesting that a prejudice – distrust of "infinity" – can be used in a positive way. And, of course, things can be different than they seem.

Again I invite you to kick around my thought that "*A cogent theory can be found which supports any set of non-contradictory postulates.*" The catch would seem to be in that adjective: "non-contradictory." I know from experience that our minds try to make sense, or pseudo-sense, out of any idea or even any piece of bigotry. I may spot an obvious contradiction right off the bat. Yet once in a while, I have found absurdity in a tale that I had accepted for years – often since childhood. We do tend to fool ourselves. What do you think? I did conclude that an ultra-high overall density (a "singularity") could not happen in a universe containing a finite amount of matter.

If my orbiting and shape-shifting torus is, in fact, a stable dynamic form, then I believe it will exist somewhere in nature. It may appear in the microcosm if not the mega-macrocosm. Our main doubts must first center on its stability. When that is settled we can go on to the Big Bang and sightline problems. If and when all such matters agree with observations, we will have a valid theory of cosmology. That theory could lead further, but it would not truly get rid of "infinity". It is always possible to ask what lies beyond. For instance: there could be any number of such universes, each totally isolated from all the others. The unanswerable abounds.

Cosmology makes a neat sort of top end for our knowledge spectrum. I suppose it takes up the largest space we are likely to fall into. But that means little of itself. Often there is a kind of serendipity to be found in working out abstract problems. That is, we may stumble onto something that is useful and closer to where we all live. On the other hand, the unknown author of the tale about princes from the land of *Serendip* gave us a handy word and a nifty thought. Perhaps that word is more important than my cosmology. Right-brained people take note. Anyway, all knowledge is useful and if my cosmology has a fatal flaw, that too is worth knowing.

For you who don't cotton to physics, may I suggest: sociology. It's more art than science in spite of the –logy suffix. How about its sub-specialty: penology? Here I'm talking about something of which I know next to nothing. But that won't stop me. It seems that prisons are the largest part of penology. They are filled, at great expense, with a significant portion of our population. They don't seem to work very well. Inmates can only associate with criminals. When an inmate is paroled, he is warned not to associate with criminals on penalty of being returned to prison. Young inmates on their first offense are mixed into the general population where they learn to embrace common criminality. Older or stronger inmates try to mentor and lead younger or weaker ones. Low IQ's try to dominate their group and higher IQ's try camouflage. A few learn things that are legal and useful outside of prison. Most do not. Cliques form along racial lines or other dubious values. Graduates often return for another go around.

Our justice system also could use improvement, and such might help relieve some prison problems. But that is not penology and not what I'm suggesting here. Recall that Gilbert and Sullivan's Mikado spent much of his time trying "to make the punishment fit the crime." Perhaps this endeavor could turn you into a king instead of a god. Maybe Mel Brooks should give it a try.

There are many good people working in penology and they have made things better. But they may be too close to their subject and things still aren't very good. I doubt that the answer lies in more new prisons or more cheerful shades of paint. Prisons were a small part of the penal system before the nineteenth century. Corporal punishment ruled in those bad old days. Stocks, flogging, torture, mutilation and death were imposed, sometimes embellished or in combination. At least they were cheap and quick. Speed and low cost are vital. Make sure that you take in <u>total</u> cost, not just current cost. Penology needs a fundamental rethink, in my opinion.

If you can come up with a truly decent penal system, you may do a lot more real good for your species than does my cosmology. If you can get it adopted, you will have to use magic and thus become a god. You will be entitled to call yourself: "Master of the Pen."

But hey, so what if you don't? I've already made you a god. If you have worked through chapters 4, 5 and 6, you have seen a universe all at once, instantaneously, without using light. That is impossible, but

GODLINESS: Chapter Nine

you will have done it: plain magic! You also will have done some lesser bits of impossible magic such as speeding up and reversing the direction of time, changing your size by 17 orders of magnitude and popping in and out of a universe. I said I would do it and so I have. Welcome to godhood!

If you have not followed through those chapters, you have not become a god. I won't know it, but you will. It seems a shame, but gods can't cheat on their initiation. Later on you can cheat as much as ever you did before. Judging from the old mythologies, you can cheat more than common people. That's a worse shame. At least try to be courteous about it.

Calculations for Chapter 5

D on't bother with this appendix unless you perversely want to catch me out. The style is dull, the math condensed. Herein are stashed most of the formulas used for Chapter 5. Symbols and values in the mks system follow:

BHIS = Black Hole in Space.

c = Relativity Constant, $2.998 \cdot 10^8$ mtr/sec.

E = Energy or Work, in joules.

e = Base of Natural Logarithms, 2.71828...

f = Force, in newtons = N = kg·mtr/sec^2.

ϕ = Gravitational Field Potential, in joule/kg.

G = Gravitational Constant, $6.67 \cdot 10^{-11}$ joule·mtr/kg^2.

GM = A Source of Excitation of Phi, in joule·mtr/kg.

joule = Unit of Work or Energy, N·mtr = kg·mtr^2/sec^2.

light-year = $9.46 \cdot 10^{15}$ mtr.

M = Large Mass, in kilograms = kg.

m = Small Mass, in kilograms = kg.

π = Euclidean Circumference/Diameter, 3.14159...

R, r = Distance, in mtr, from center of GM.

ω = Angular Velocity, omega, in radian/sec.

The two definitions of ϕ in Relativity Theory are: $\phi = -GM/r$ and $\phi = -\omega^2 r^2/2$. The energy equivalence of matter is given by E = mc^2. From these definitions, one can generalize that a mass, m_ϕ, is decreased by the presence of a gravitational field potential relative to the same mass, m_0, where $\phi = 0$. The relationship is:

$$m_\phi = \frac{m_o}{1 + \dfrac{-\phi}{c^2}} \qquad (1)$$

From Newton's force of gravitational attraction, it follows that:

$$f = \frac{GMm_0}{r^2} = GMm_\phi \frac{1 + \dfrac{-\phi}{c^2}}{r^2}; \ let \ m_\phi = 1, \ then$$

$$-\phi = \int_r^\infty GM \frac{1 + \dfrac{-\phi}{c^2}}{r^2} \, dr. \qquad From\ calculus,$$

$$\int a(u \pm v)dx = a\left[\int u\,dx \pm \int v\,dx\right] and \int u^n du = \frac{u^{n+1}}{n+1},$$

$where \ n \neq -1. \ First\ approximat\ ion, \ -\phi = \dfrac{GM}{r}.$

$$So \ -\phi = \frac{GM}{r} + \frac{(GM)^2}{2c^2 r^2} + \frac{(GM)^3}{6c^4 r^3} + ... \ Let \ x = \frac{GM}{c^2 r},$$

$$so \frac{-\phi}{c^2} + 1 = 1 + x + \frac{x^2}{2!} + \frac{x^3}{3!} + ... \ Maclaurin's \ series$$

$$resolves \ to: \ -\phi = c^2\left(e^{\frac{GM}{c^2 r}} - 1\right) \qquad (2)$$

Equation (3) is developed in a similar fashion, still for $m_\phi = 1$:

$$-\phi = c^2\left(e^{\frac{\omega^2 r^2}{2c^2}} - 1\right) \qquad (3)$$

A stable circular orbit can exist when ϕ from Equation (2) equals ϕ from Equation (3), in which case:

$$\frac{GM}{c^2 r} = \frac{\omega^2 r^2}{2c^2}, \; next \quad \frac{GM}{r} = \frac{\omega^2 r^2}{2} = \frac{v^2}{2}.$$

So BHIS conditions are met when $\dfrac{GM}{c^2 r} = \dfrac{1}{2}.$

Substitution in Equation (2) *or* (3) *gives* :

$$-\phi = c^2\left(\sqrt{e} - 1\right) = c^2(0.64872\ldots) \qquad (4)$$

The $-\phi$ calculated from Equation (4) refers to a situation where orbiting velocity, v_o, is equal to c and, hence, is the limiting maximum possible velocity. It defines a BHIS that will not permit any form of energy to rise above its event horizon.

It is interesting to try to describe conditions within and near such a BHIS. From the viewpoint of an observer in a Galilean reference body where $-\phi$ is negligible (similar to our own reality of astronomical observations), $-\phi$ is zero at the center of this BHIS and is $5.831 \cdot 10^{16}$ joule/kg at distance $r = GM/-\phi$, and all of the exciting source, GM, is contained within the distance r. Although this BHIS has a definite center and a definite radius, it is not a Euclidean sphere since the circumference of its great circle $\neq 2\pi r$, its volume $\neq (4/3)\pi r^3$ and its area $\neq 4\pi r^2$. In fact, its area is infinite at distance r where $-\phi = c^2 \left(\sqrt{e} - 1 \right)$ although it is finite at any other distance or where $-\phi$ has a lesser value. Hence, the concept of density loses meaning at position r; it approaches zero, being a mass divided by infinity! Also, a unit mass at position r has much less potential energy than when at the center.

The observer, thinking of Boyle's law, tends to believe that all of the mass is located in a very thin "shell" of radius r, and that all of the interior of the BHIS is nearly empty space where $-\phi$ is near zero and blackbody temperature is near absolute zero.

[An English friend once got steamed up over my use of the phrase, "blackbody temperature." Note that it does not particularly involve ladies of African descent. It refers to the equilibrium temperature of an object that is a perfect receiver and emitter of radiant energy.]

This is a good example of the weirdness that can arise from mixing infinity with material things. And yet it does define a sort of limiting maximum value for $-\phi$, analogous to c being the limiting maximum value for any velocity.

Fortunately, the idea that no form of energy can rise above the event horizon of a BHIS is not necessary to the BHIS concept; that concept only requires that no form of energy can permanently escape from a BHIS – that is, that escape velocity be unattainable.

Escape velocity can be easily determined by equating negative gravitational field potential, $-\phi$, with specific kinetic energy,

$(1/2)mv^2/m$. It follows: $v_e = \sqrt{-2\phi} = \sqrt{\dfrac{2GM}{r}}$.

For example, at the surface of the earth: GM/r^2 = acceleration of gravity = 9.81 mtr/sec^2, $r = 6.63\cdot10^6$ mtr and $M = 6.47\cdot10^{24}$ kg, so $-\phi = 6.50\cdot10^7$ joule/kg – thus, escape velocity = 11,400 mtr/sec or 25,500 miles/hr. The earth's rotation and its atmospheric resistance will modify this escape velocity relative to a surface firing point.

To an observer in a region of negligible value of $-\phi$, the velocity of electromagnetic radiation propagation in a region where $-\phi$ is significant is v_r, and $v_r = c\sqrt{1 - \dfrac{-\phi}{c^2\left(\sqrt{e}-1\right)}}$. For the newly defined BHIS, this must be infinitesimally smaller than the escape velocity. The energy difference that a unit mass shows at rest and at this velocity is the required $-\phi$. It follows:

$$\frac{\phi}{c^2} = 1 - \frac{1}{\sqrt{1 - \frac{v_r^2}{c^2}}} = 1 - \frac{1}{\sqrt{1 - \frac{-\phi}{c^2(\sqrt{e}-1)}}} \cdot$$ Change

to $\dfrac{\phi^2}{c^4} - \dfrac{2\phi}{c^2} + 1 = \dfrac{c^2(\sqrt{e}-1)}{-\phi}$, and with a little more algebraic manipulation:

$$\phi^3 - 2c^2\phi^2 + c^4\phi + c^6(\sqrt{e}-1) = 0 \qquad (5)$$

The roots of this cubic equation are:
$\phi = -0.353902354...c^2$ and two complex roots. At this very high level of $-\phi$,
$\dfrac{v_r}{c} = \sqrt{1 - \dfrac{0.35390}{0.64872}} = \sqrt{0.45446} = 0.67414$ and the length of a measuring rod perpendicular to the radius relative to the same rod parallel to the radius, or at a place where $-\phi = 0$, is also 0.67414, as is the rate of a clock. Hence, the surface area of such an object equals $4\pi r^2/0.45446$, and this value of $-\phi$, $3.181 \cdot 10^{16}$ joule/kg, is the minimum that can define a BHIS. If this "minimal" BHIS were rotating, then radiation leaving its surface in an easterly direction would, in fact, escape.

* * *

The next major item concerns the $-\phi$ topography of a torus. A Euclidean torus is a surface generated by a circle of radius r, rotated about an axis in its plane that does not intersect the circle. The center of this smaller circle is at distance R from this axis. The definition requires that R be equal to or greater than r. For this shape: volume = $2\pi^2 Rr^2$, and surface area = $4\pi^2 Rr$. The definition of a toroid is similar, except that its generating shape is not a circle, but may be any defined shape. If R/r is very large, then the centerline (which I here call the X-axis) approaches a straight line and a local section is in the form of a tube. This centerline of mass distribution is such that ρ_L = "line density" in kg/mtr. The gravitational force (f) acting on a mass (m_ϕ) located in Cartesian co-ordinates where y = r, x = 0 is found from: $f = \dfrac{GM_E m_\phi}{r^2}$. The equivalent mass, M_E, might be found from:

$$M_E = 2\rho_L \int_0^\infty \frac{r}{x^2+1}\,dx = 2\rho_L r \int \frac{1}{x^2+1}\,dx.$$

Rearranging and integrating gives:

$$\frac{M_E}{2\rho_L r} = \int \frac{1}{x^2+1}\,dx = \tan^{-1}(x) = \frac{\pi}{2}. \text{ Let } m_\phi = 1,$$

and then: $f_s = \dfrac{G(\pi\rho_L r)}{r^2} = \dfrac{G(\pi\rho_L)}{r}.$ So,

$f_s r = -\phi = G(\pi\rho_L)$, and r is gone from the equation for $-\phi$! This seeming oddity comes from describing the center of mass as an infinitely long straight line fixed in empty space.

To approach things differently, look at the whole centerline of mass as a circle (parallel of latitude) of radius R, lying on the surface of Yaul's Sphere, with its radius R_S = $1.224 \cdot 10^{26}$ mtr. Imagine an outer sphere of radius R_o concentric with Yaul's Sphere. Let $r_o = R_o - R_S$. The $-\phi$ at the outer end of R_o would be the sum of the $-\phi$ from the total mass of the universe, M_U, vectored from the center of Yaul's Sphere plus the $-\phi$ vectored directly from the closer portions of M_U, call them M_E. Then imagine an inner sphere of radius R_i concentric with Yaul's Sphere. Let $r_i = R_S - R_i$. The $-\phi$ at the outer end of R_i would be the $-\phi$ vectored directly from the closer portions of M_U, M_E, less the $-\phi$ from the total mass of the universe, M_U, vectored from the center of Yaul's Sphere.

Any $-\phi$ that is of interest in defining our toroid will be in the range of, perhaps, 5% greater than that of a minimal BHIS to just less than that of a maximal BHIS, say between $3.35 \cdot 10^{16}$ and $5.80 \cdot 10^{16}$ joule/kg. A near minimal BHIS $-\phi$ may be suitable for the outer end of R_o as there are no other universes nearby and whatever radiation may leave this point will eventually return to the toroid. As $-\phi$ at the surface of the Sphere is $2.234 \cdot 10^{16}$ joule/kg, r_o will be increased by this general effect, and r_i will be similarly decreased. A value nearer the maximal $-\phi$ for a BHIS, say up to $5.43 \cdot 10^{16}$ joule/kg as calculated from equation (7) below, could be suitable for latitudes greater than zero.

In general, let's assume that the apex of our toroid (outer end of R_o) will be at the point where $-\phi$ from the angular velocity equals the sum of $-\phi$ from both the Sphere's center and $-\phi$ from the nearby equivalent mass, M_E. An object, perhaps a star, at this outer place could move along in sync with the centerline without any special tendency to move ahead or behind, up or down. This is a handy, if not requisite, assumption.

*

Consider the special case at Lat 0°, the Equator, where $R = R_S = 1.224 \cdot 10^{26}$ mtr, and $\rho_L = 4.19 \cdot 10^{25}$ kg/mtr. The outermost end of vectored R_o must move along in sync with the centerline of mass. Consequently, it will have the same angular velocity, ω, as that centerline. It follows that: $-\phi = \dfrac{\omega^2 R^2}{2}$,

$$R^2 = \frac{-\phi \times 2}{\omega^2}, \text{ so } \frac{R_o}{R_S} = \sqrt{\frac{3.35}{2.234}} = 1.22456 \quad (6a)$$

Hence, $R_o = 1.22456 \cdot R_S = 1.4989 \cdot 10^{26}$ and $r_o = 0.22456 \cdot R_S = 2.749 \cdot 10^{25}$ mtr (2.91 billion light-years). Although it is easy to think of this as a maximum value for r_o, it must, in fact, be the only value.

GODLINESS: Appendix A

At this outside point, $-\phi$ from the Sphere's center contributes $1.824 \cdot 10^{16}$ joule/kg and $-\phi$ from the nearby mass must add another $(3.35 - 1.824 = 1.526) \cdot 10^{16}$ joule/kg. This implies that the nearby M_E is $6.287 \cdot 10^{51}$ kg. Since the total mass, M_U, is $4.10 \cdot 10^{52}$ kg, the nearby M_E is 15.3% of that total.

Next, find the radius, R_i, of a sphere concentric with and inside of Yaul's Sphere such that the distance, r_i, from the formula $r_i = R_S - R_i$, is the distance from the centerline of mass to where $-\phi = 3.35 \cdot 10^{16}$ joule/kg. Note that this $-\phi$ is the net result of two gravitational field potentials in direct opposition. Set up an equation for R_i:

Let $-\phi = a = 3.35 \cdot 10^{16}$, $GM_U = b = 2.7347 \cdot 10^{42}$,

$GM_E = c = 4.1937 \cdot 10^{41}$, and $R_S = d = 1.224 \cdot 10^{26}$.

$-\phi = \dfrac{GM_E}{R_S - R_i} - \dfrac{GM_U}{R_i}$, so $ad - aR_i = c - \dfrac{bd}{R_i} + b$, and

$$\therefore aR_i^{\,2} - (ad - b - c)R_i - bd = 0. \qquad (6b)$$

The usual form is: $aR_i^{\,2} + bR_i + c = 0$; where $a = 3.35 \cdot 10^{16}, b = -9.463 \cdot 10^{41}, c = -3.3473 \cdot 10^{68}$.

108

The useful root is $R_i = 1.15076 \cdot 10^{26}$ mtr. Hence $r_i = 7.324 \cdot 10^{24}$ mtr (774 million light-years).

Last, find the distances on the surface of Yaul's Sphere, r_f and r_b, forward from the centerline of mass and behind it, where $-\phi = 3.35 \cdot 10^{16}$ joule/kg and $(r_f + r_b)/2 = r_{fb}$. There is no reinforcement due to another gravitational field potential at a right angle. Using the same sort of calculation as above,

$$-\phi = \frac{GM_E}{r_{fb}} \quad \text{and} \quad r_{fb} = \frac{4.1937 \cdot 10^{41}}{3.35 \cdot 10^{16}} = 1.252 \cdot 10^{25}.$$

This value is in meters and is equivalent to 1.323 billion light-years.

The cross-section of this toroid is a form made of two ellipses. The inner ellipse has its long axis lying along the surface of Yaul's Sphere while the outer ellipse is perpendicular to that surface. Curvature of the surface adds another complication. I want to calculate the volume of this toroid and will simplify its cross-section to a circle whose radius is the average of the four determined: $1.496 \cdot 10^{25}$ mtr. [The shape center (not mass center) of this circle is $1.008 \cdot 10^{25}$ mtr (8.2%) above the surface of Yaul's Sphere. But I decided not to make minor adjustments at this early stage of development.] So, I set up a spreadsheet to integrate the specific volume of a one-meter slice of the toroid and multiply it by the circumference of a great circle of the Sphere, $9.792 \cdot 10^{26}$ mtr. The General Relativity volume of this toroid at Lat $0°$ turned out to be $1.027 \cdot 10^{78}$ mtr^3.

*

Consider the other special case at Lat $90°$, the Pole, where R = 0 and the centerline of mass is a point of equivalent mass: $M_U = 4.10 \cdot 10^{52}$ kg. R_o would be greater here than at the Equator.

First, find the intersection circle for a $-\phi$ of $5.43 \cdot 10^{16}$ joule/kg: $(r_f + r_b)/2 = r_{fb} = 6.67 \cdot 10^{-11} \cdot 4.10 \cdot 10^{52} / 5.43 \cdot 10^{16} = 5.036 \cdot 10^{25}$ mtr (5.32 billion light-years).

Next, an equation can be set up for the inside radius, R_i:

Let $-\phi = a = 5.43 \cdot 10^{16}$, $GM_U = b = 6.67 \cdot 10^{-11} \cdot 4.10 \cdot 10^{52}$,

$R_S = c = 1.224 \cdot 10^{26}$. $-\phi = \dfrac{GM_U}{R_S - R_i} - \dfrac{GM_U}{R_i}$, so

$ac - aR_i = b - \dfrac{bc}{R_i} + b$, and

$$\therefore aR_i^{\,2} - (ac - 2b)R_i - bc = 0. \qquad (6c)$$

In usual form: $aR_i^{\,2} + bR_i + c = 0$; where $a = 5.43 \cdot 10^{16}, b = -1.1769 \cdot 10^{42}, c = -3.347 \cdot 10^{68}$. There is one useful root: $R_i = 9.009 \cdot 10^{25}$ mtr. So $r_i = 3.231 \cdot 10^{25}$ mtr (3.42 billion light-years).

R_o can be found to satisfy both the angular velocity formula and the $G \cdot M_U$ formula, as at Lat $0°$. It must give a $-\phi$ in the range between 3.35 and 5.80 $\cdot 10^{16}$ joule/kg.

Develop an equation: *Since* $\; -\phi = \dfrac{\omega^2 R^2}{2}$,

Let $\;\dfrac{\omega^2}{2} = a = \dfrac{2.234 \cdot 10^{16}}{\left(1.224 \cdot 10^{26}\right)^2} = 1.49115 \cdot 10^{-36}$.

$\therefore -\phi = aR_o^{\,2}$. *Also,* $-\phi = \dfrac{GM_U}{R_o} + \dfrac{GM_U}{R_o - R_S}$. *Let*

$GM_U = b = 6.67 \cdot 10^{-11} \cdot 4.10 \cdot 10^{52} = 2.7347 \cdot 10^{42}$.

Let $R_S = c = 1.224 \cdot 10^{26}$. $\therefore aR_o^2 = \dfrac{b}{R_o} + \dfrac{b}{R_o - c}$. So,

$$aR_o^3 - acR_o^2 = b - \dfrac{bc}{R_o} + b, \quad -\dfrac{bc}{R_o} = aR_o^3 - acR_o^2 - 2b, \text{ or}$$

$$\dfrac{aR_o^4}{bc} - \dfrac{aR_o^3}{b} - \dfrac{2R_o}{c} + 1 = 0. \qquad (7)$$

The easiest way to find a real root of this quartic equation without any special software is by trial and error in a spreadsheet program. I started with $R_o = 4 \cdot 10^{26}$ and worked down to bracket the zero to four decimal places of R_o. The root is: $R_o = 1.9082 \cdot 10^{26}$ mtr. So $r_o = 6.842 \cdot 10^{25}$ mtr = 7.23 billion light-years, and $-\phi = 5.430 \cdot 10^{16}$ joule/kg. That, of course, is the greater value of $-\phi$ that I used in equation (6b).

I want to calculate the General Relativity volume of this shape as I did for Lat 0°. Its average radius is $5.036 \cdot 10^{25}$ mtr and its shape can be taken as spherical. Its shape center is $1.805 \cdot 10^{25}$ mtr above the surface of Yaul's Sphere, although this does not affect its volume. From my integrating spreadsheet, its General Relativity volume works out to be $1.917 \cdot 10^{78}$ mtr^3 – almost twice the volume at the Equator.

*

As the toroid approaches the Pole, it starts to collide with itself around Lat 72° and will continue to display an increasing frequency of collisions until its mass center becomes a point at the Pole. Thereafter, collisions will become less frequent until they stop entirely near Lat 72°. These fireworks carry on for about ten billion years.

Since R_o at the Pole is greater than R_o at the Equator (ratio of 1.9082/1.4989), the imaginary outside sphere that I referred to turns out to be an ellipsoid about 27% out of round. Its long axis coincides with the axis of Yaul's Sphere. The imaginary inside sphere is a shape (ratio of 0.9009/1.1508) nearly 28% out of round whose short axis coincides with the axis of Yaul's Sphere. The dimensions of the toroid at intermediate latitudes can be found from the geometry of the outer ellipse and the general relationships already used.

*

Analytic geometry, for an ellipse centered on Cartesian coordinates so that its two axes coincide with the X- and Y-axes, yields the equation: $\frac{x^2}{a^2}+\frac{y^2}{b^2}=1$, where a and b are the semi-axes. (For a circle, a = b = radius.) Let the X-axis coincide with the axis of Yaul's Sphere, and the Y-axis intersect the Sphere on its Equator at the 0° and 180° meridians of longitude.

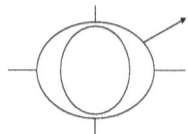

Finally, take the general case, at Lat 45°N. Let (90° – Lat) = α; then y = (tan α)x = x at Lat 45°. The equation becomes: $x^2\left(\frac{1}{a^2}+\frac{(\tan\alpha)^2}{b^2}\right)=1$. For the outer ellipse: a_o = 1.9082·10²⁶ mtr and b_o = 1.4989·10²⁶ mtr. So, $x_o{}^2$ = $y_o{}^2$ = 1.3894·10⁵². Hence, R_o = 1.6670·10²⁶ mtr and r_o = 4.430·10²⁵ mtr

(4.68 billion light-years). [For an inner ellipse: a_i = 9.009·10^{25} mtr and b_i = 1.1508·10^{26} mtr. So, x^2 = y^2 = 5.0322·10^{51}. Hence, R_i = 1.00322·10^{26} mtr and r_i = 2.2079·10^{25} mtr (2.33 billion light-years) if the inner shape were an ellipse.]

−ϕ is found by a variant of equation (6a):

$-\phi = 2.234 \cdot 10^{16} \left({R_o}^2 \middle/ {R_S}^2 \right)$. This works out to

be −ϕ = 4.077·10^{16} joule/kg. At the outer end of R_o, −ϕ from the Sphere's center is 1.6405·10^{16} joule/kg and −ϕ from the nearby mass must be (4.077 − 1.6405 = 2.4365) ·10^{16} joule/kg. This implies that the nearby M_E is 1.618·10^{52} kg. The total mass, M_U, is 4.10·10^{52} kg, so the nearby M_E is 39.5% of that total.

[M_E might also be found from the equation used to develop (6b):

Let $-\phi = 4.077 \cdot 10^{16}$, $GM_U = 2.7347 \cdot 10^{42}$, $R_S = 1.224 \cdot 10^{26}$, $R_i = 1.00322 \cdot 10^{26}$, $G = 6.67 \cdot 10^{-11}$. Since $-\phi = \dfrac{GM_E}{R_S - R_i} - \dfrac{GM_U}{R_i}$, then

$$M_E = \frac{R_S - R_i}{G}\left(-\phi + \frac{GM_U}{R_i}\right).$$

Here $M_E = 2.2518 \cdot 10^{52}$ kg. The difference between this M_E and the previous M_E is due to the inner shape not being an ellipse.]

Now, find R_i such that the distance, r_i, from the formula, $r_i = R_S - R_i$, is the distance from the centerline of mass to where $-\phi = 4.077 \cdot 10^{16}$ joule/kg. Note that this $-\phi$ is the net result of two gravitational field potentials in direct opposition. Make use of equation (6b):

Let $-\phi = a = 4.077 \cdot 10^{16}$, $GM_U = b = 2.7347 \cdot 10^{42}$, $GM_E = c = 1.0792 \cdot 10^{42}$, and $R_S = d = 1.224 \cdot 10^{26}$.

Since $-\phi = \dfrac{GM_E}{R_S - R_i} - \dfrac{GM_U}{R_i}$, then $ad - aR_i = c - \dfrac{bd}{R_i} + b$,

and $\therefore aR_i^2 - (ad - b - c)R_i - bd = 0$.

The usual form is: $aR_i^2 + bR_i + c = 0$; where $a = 4.077 \cdot 10^{16}, b = -1.1763 \cdot 10^{42}, c = -3.3473 \cdot 10^{68}$. The useful root is $R_i = 1.06177 \cdot 10^{26}$ mtr. Hence $r_i = 1.6223 \cdot 10^{25}$ mtr (1.715 billion light-years).

Find the distance, r_{fb}, on the surface of Yaul's Sphere forward from the centerline of mass and behind it, where $-\phi = 4.077 \cdot 10^{16}$ joule/kg. There is no reinforcement due to another gravitational field potential at a right angle. So, $-\phi = \dfrac{GM_E}{r_{fb}}$, $\quad r_{fb} = \dfrac{1.0792 10^{42}}{4.077 \cdot 10^{6}} = 2.647 \cdot 10^{25}$.

This value is in meters and is equivalent to 2.80 billion light-years.

I want to calculate volume for this toroid and will simplify its cross-section to a circle whose radius is the average of the four determined: $2.837 \cdot 10^{25}$ mtr. I set up a spreadsheet to integrate the specific volume of a one-meter slice of the toroid and multiplied that by the circumference of a great circle of the Sphere ($9.792 \cdot 10^{26}$ mtr) times the cosine of the Latitude (0.7071). This General Relativity volume of the toroid at Lat 45° came out as $3.021 \cdot 10^{78}$ mtr^3 – about half again its volume at the Equator or the Pole. Don't forget that the apparent volume to observers within the Monodim's Universe is much greater.

* * *

Math for this $-\phi$ topography is far from rigorous. I had some conceptual difficulty in summing up $-\phi$ from two different source locations. Values chosen for $-\phi$, particularly $3.35 \cdot 10^{16}$ joule/kg, are rather arbitrary. As has been mentioned, values for $-\phi$ should be worked up from careful detailing of matter distribution over the General Relativity volumes. The effect of curvature of the surface of Yaul's Sphere on the shape of cross-sectional areas has been ignored. And visual distances are still to be determined for observers within the Monodim's Universe. Elaborate calculation of matter distribution within the toroid is needed for various reasons. This will be left for some future amusement. Newer explicit math has been done in Chapter 8, written after this Appendix.

About the Author

Charles Heath claims Mystic, Connecticut as his hometown. He left the States to live in South America with a freshly minted degree in Chemical Engineering. After a stint with the Marines at the time of the Korean War, he returned to South America, married and briefly worked with Exxon's then new grass-roots refinery at Cartagena. He later moved to the Caribbean island of Barbados.

After more than three decades spent in Barbados, he retired to become a cruising liveaboard on the yacht *Cosmos* – sailing the Caribbean for a few years. In 1998, he settled in Houston where two of his three daughters reside.

Charles Heath has always been self-employed, except for those two years with Exxon. He is interested in flying, films, writing and music.

www.ingramcontent.com/pod-product-compliance
Lightning Source LLC
Chambersburg PA
CBHW022007170526
45157CB00003B/1180